T0180201

Environmental Footprints and Eco-design of Products and Processes

Series editor

Subramanian Senthilkannan Muthu, SgT Group and API,
Hong Kong, Hong Kong

This series aims to broadly cover all the aspects related to environmental assessment of products, development of environmental and ecological indicators and eco-design of various products and processes. Below are the areas fall under the aims and scope of this series, but not limited to: Environmental Life Cycle Assessment; Social Life Cycle Assessment; Organizational and Product Carbon Footprints; Ecological, Energy and Water Footprints; Life cycle costing; Environmental and sustainable indicators; Environmental impact assessment methods and tools; Eco-design (sustainable design) aspects and tools; Biodegradation studies; Recycling; Solid waste management; Environmental and social audits; Green Purchasing and tools; Product environmental footprints; Environmental management standards and regulations; Eco-labels; Green Claims and green washing; Assessment of sustainability aspects.

More information about this series at http://www.springer.com/series/13340

Subramanian Senthilkannan Muthu
Editor

Environmental Water Footprints

Energy and Building Sectors

 Springer

Editor
Subramanian Senthilkannan Muthu
SgT Group and API
Hong Kong, Kowloon, Hong Kong

ISSN 2345-7651   ISSN 2345-766X   (electronic)
Environmental Footprints and Eco-design of Products and Processes
ISBN 978-981-13-4805-1   ISBN 978-981-13-2739-1   (eBook)
https://doi.org/10.1007/978-981-13-2739-1

This Springer imprint is published by the registered company Springer Nature Singapore Pte Ltd.
The registered company address is: 152 Beach Road, #21-01/04 Gateway East, Singapore 189721,
Singapore

This book is dedicated to:
The lotus feet of my beloved Lord
Pazhaniandavar
My beloved late Father
My beloved Mother
My beloved Wife Karpagam and Daughters-
Anu and Karthika
My beloved Brother
Everyone working in the Energy & Building
sectors to make it Environmentally
Sustainable

Contents

The Water Footprint in Bioenergy—A Comparison of Four Biomass Sources to Produce Biofuels in Argentina

P. Araujo, A. P. Arena, B. Civit, S. Curadelli, S. Feldman, E. Jozami, F. Mele, R. Piastrellini and J. Silva Colomer

Abstract In Argentina, the use of biomass for the production of biofuels capable of replacing fossil fuels has aroused great expectation. In the last decade, the production of biodiesel has tripled, and the rising trend continues supported by national legislation that soon will increase the blend mandates up to B20 and E25 (being now B10 and E12). However, there could be some environmental concerns associated with land use and especially with water use. In this chapter, we calculate the volume of water used and consumed in the production of rapeseed (*Brassica napus*) and soybean (*Glycine max*) for biodiesel, and the production of sugarcane (*Saccharum officinarum*) and cordgrass (*Spartina argentinensis*) for bioethanol in different regions of the country. The water footprint, as defined by the Water Footprint Network, is used as an indicator of water resources appropriation, and the ISO approach

P. Araujo · F. Mele
Departamento de Ingeniería de Procesos y Gestión Industrial, Universidad Nacional de Tucumán, Av. Independencia 1800, T4002BLR San Miguel de Tucumán, Argentina

A. P. Arena · B. Civit (✉) · F. Mele
Consejo Nacional de Investigaciones Científicas y Técnicas, Godoy Cruz 2290, C1425FQB CABA, Argentina
e-mail: bcivit@mendoza-conicet.gob.ar; barbara.civit@gmail.com

A. P. Arena · B. Civit · S. Curadelli · R. Piastrellini
Facultad Regional Mendoza, Universidad Tecnológica Nacional, Cnel. Rodríguez 273, 5500 Mendoza, Argentina

E. Jozami
Facultad de Ciencias Agrarias, Universidad Nacional de Rosario, Campo Experimental Villarino C.C. 14, S2125ZAA Zavalla, Santa Fe, Argentina

J. Silva Colomer
Instituto Nacional de Tecnología Agropecuaria, EEA Cuyo, San Martin 3853, 5507 Luján de Cuyo, Mendoza, Argentina

S. Feldman
Instituto de Investigaciones en Ciencias Agrarias, CONICET, Facultad de Ciencias Agrarias, Universidad Nacional de Rosario, Campo Experimental Villarino C.C. 14, S2125ZAA Zavalla, Santa Fe, Argentina

© Springer Nature Singapore Pte Ltd. 2019
S. S. Muthu (ed.), *Environmental Water Footprints*, Environmental Footprints and Eco-design of Products and Processes,
https://doi.org/10.1007/978-981-13-2739-1_1

is followed to assess the impacts associated with the use of water. The volume of water associated with the production of cordgrass is lower than figures obtained by traditional sources of biomass in Argentina (soybean and sugarcane). Soybean is produced in the Pampean Region and it is recommended to optimize the management of water resources in that region to minimize competition with food products while on the opposite, there is the case of *Spartina* that is a native grass growing naturally in the Chaco Region and it uses water that does not compete with food or livestock feed. On the other hand, rapeseed has a high water footprint mainly as a consequence of the site where it is produced. For instance, considering the environmental fragility, it is recommended to avoid the production of biomass destined to bioenergy in the arid zones of the country. Therefore, our findings show that results are more dependent on the region where each biomass is grown, than on management practices or the amount and type of chemical inputs (fertilizers, pesticides). Further work, such as the accounting of water along the industrial phase of the biofuels, is needed to have the full picture of the water consumption issue in bioenergy production.

Keywords Biomass sources · Biodiesel · Bioethanol · Water use · Scarcity Water footprint

This chapter deals with the use of water in the production of biomass with energy purpose and is organized in five main sections. Section 1 introduces the reader to the global and regional situation related to bioenergy and biofuels production and also, discusses the environmental issues derived from their uses. In Sect. 2, general concepts of Water Footprint in the field of bioenergy production are exposed. After that, in Sect. 3, the functional unit, the system boundaries and the four case studies are described. Section 4 outlines the results obtained for each case study and discusses them together. Finally, in the last section, authors draw conclusions and make recommendations for future work that complements the subject addressed.

1 Introduction

Renewable energies are continuously expanding throughout the world, not only because of the technological advances achieved in terms of production and storage, but also because of the implementation of promotion policies for their development, by the environment commitments assumed, and by the growing interest in accessing to alternative energy sources (REN 21 2017).

Within renewable energies, bioenergy is the one that contributes most to the global energy supply, playing an important role in the provision of heat, generation of electricity and also in the transportation sector through the use of biofuels.

In the present, many countries, especially in America and Europe, include biofuels in their energy matrix and have promotion policies for their production and use, such as subsidies and mandatory regulations for blending with fossil fuels for transportation. However, there is a substantial debate about the real environmental sustainability of biofuel production, closely associated with the use of resources, especially land and water.

1.1　Bioenergy and Energy Crops

Biomass is the organic matter originated in a biological process, either spontaneous or incited. It is therefore the biodegradable fraction of the products and residues of agriculture, livestock, afforestation and its associated industries, municipal waste, and industrial waste. The term contemplates organic matter formed by biological means in the immediate past or in the products derived from it; for that reason it excludes the fossil formations.

Biomass can be used to generate energy, called bioenergy. It is considered a renewable source because its energy content comes, ultimately, from the solar energy fixed by plants and a few microorganisms during the photosynthetic process. Through photosynthesis, chlorophyll captures solar energy, and turns carbon dioxide from air and soil water into carbohydrates, to form organic matter. When the organic matter is burned, the carbohydrate molecules are broken, releasing the energy contained in them. In this way, the biomass behaves like a "battery" that stores solar energy.

Bioenergy is the oldest form of energy and has contributed the most to technological development. It has had various uses, from cooking and heating homes to powering steam engines. The remarkable increase in the energy demand of the mid-eighteenth century led to the use of more intensive energy sources in terms of fuel power, including coal. From that moment, the use of biomass decreased to historical lows that coincided with the massive use of petroleum products. Despite this, biomass still continues to play a prominent role as an energy source in different industrial and domestic applications.

There are different types of biomass that can be used to produce energy. The most widespread typology is based on the process that originated it (spontaneous or induced processes). According to this classification, four types of biomass are distinguished:

(i)　Natural biomass, produced in natural ecosystems without human intervention.

(ii)　Dry residual biomass, generated as a result of human activities in the form of agricultural, livestock, forestry, industrial and municipal waste; with moisture content less than 60%.

(iii)　Wet residual biomass, generated as a result of human activities in the form of agricultural, livestock, forestry, industrial and municipal waste; with moisture content equal to or greater than 60%.

(iv)　Energy crops, herbaceous or woody, produced specifically for their energy value, or intended for human food or animal feed and that can also be used for energy purposes (Piastrellini 2015).

According to its final use, energy crops can be classified into:

(iv-a) Oil seed crops, with high oil content convertible to esters. Examples: soybean (*Glycine max*), rapeseed (*Brassica napus*), sunflower (*Helianthus annuus*), jatropha (*Jatropha curcas*), and castor oil plant (*Ricinus communis*).
(iv-b) Alcoholigenic crops, with high content of fermentable sugars. Examples: sugarcane (*Saccharum officinarum*), and corn (*Zea mays*).
(iv-c) Lignocellulosic crops, with high content of lignin and cellulose. Examples: poplar (*Populus* sp.), willow (*Salix* sp.), eucalyptus (*Eucalyptus* sp.).

Oil seed crops and alcoholigenic crops are the raw material par excellence for biofuels production, being corn, wheat, sugarcane, rapeseed (*Brassica napus*), soybeans, and sunflower the largest worldwide exponents.

1.2   Biofuels for Transportation

Fuels derived from biomass or its metabolic waste and that can be used in any energy application, whether thermal, electrical or mechanical, to feed boilers and internal combustion engines are called "biofuels". Depending on their physical condition, biofuels can be solid (firewood, charcoal, pellets, briquettes, etc.), liquid (biodiesel and bioethanol) or gaseous (biogas).

Liquid biofuels are used mainly in the transportation sector, either in its pure state or in mixtures with fossil fuels. Biodiesel is usually mixed in different proportions with low-sulfur diesel. The biodiesel-petrodiesel mixtures are named with the notation "BXX", where "XX" represents the volumetric percentage of biofuel in the mixture. For example: "B20" indicates a mixture with 20% biodiesel and 80% diesel of fossil origin. Bioethanol can be mixed with gasoline to achieve improvements in oxygenation and increase the octane level. The bioethanol-gasoline mixtures are commonly known as "gasohol" or "alconafta", and are identified with the notation "EXX", with a similar criteria to the notation "BXX". Mixtures of up to "E25" do not require modifications to the current vehicles engines, while higher biofuel concentrations such as "E85" imply the use of flexible fuel vehicles (FFV).

In terms of combustion efficiency, more biofuel than fossil fuel is required to achieve the same amount of energy. In general terms, approximately 1.2 kg of biodiesel and 1.6 kg of bioethanol are required to replace 1 kg of diesel and 1 kg of gasoline, respectively. This is because biofuels have lower calorific value than fossil fuels. There are two parameters associated with the calorific value of a fuel:

(i) Higher heating value (HHV): is the total heat evolved in the complete combustion of the fuel when the water vapor originating in the combustion condenses. Counts the heat released in this phase change.
(ii) Lower heating value (LHV): is the total heat evolved in the complete combustion of a fuel without counting the fraction corresponding to the latent heat of the

combustion water vapor, since it is expelled in the form of vapor and is not it produces phase change. It is important to note that the LHV is the parameter that must be used to estimate the amount of energy that can be generated from a biofuel, since this parameter indicates its net caloric power.

1.3 Biomass to Biofuels Conversion Technologies

1.3.1 Biomass Treatments

Previous to biofuels production, biomass is collected and subjected to several treatments in order to achieve the appropriate size and moisture content for processing, such as mechanical crushing, drying (natural or forced), grinding, sieving and densification.

Once the biomass reaches the appropriate size and moisture content, it is subjected to different energy conversion treatments: the dry biomass (with a moisture content less than or equal to 60%) is treated by thermochemical processes; wet biomass (with moisture content greater than 60%) is subjected to biochemical processes; and biomass with a high content of fats and oils is treated by physicochemical processes.

In thermochemical processes, biomass is subjected to high temperatures, obtaining heat or secondary products in the form of solid, liquids or gaseous fuels, such as charcoal, bio-oil and syngas.

The biochemical processes are based on the biomass degradation by the action of enzymes provided by microorganisms. They can be divided into two large groups: (i) anaerobic processes, which occur in the absence of air like methane fermentation that originates biogas; and (ii) anoxic processes, like alcoholic fermentation, through which bioethanol is produced.

In alcoholic fermentation, monosaccharides (glucose, and fructose) are converted to bioethanol (or ethyl alcohol) by means of yeasts, releasing CO_2. Currently, almost all the ethanol produced in industrial facilities is obtained using the yeast *Saccharomyces cerevisiae* in its commercial form (Sánchez Godoy 2012). Many times it is necessary to carry out processes prior to fermentation. These processes vary according to the biomass chemical composition: (i) if the biomass has high starch content (as in wheat, rice, corn, barley, rye, oats, potatoes, sweet potatoes and cassava), it is subjected to hydrolysis enzymatic; (ii) if the biomass has high cellulose and hemicellulose contents (such as cotton, wood and cereal straw), a chemical hydrolysis is carried out; and (iii) if the biomass has high sugars concentration (as in sugarcane, beet, sweet sorghum, fruits, etc.), no pre-fermentation treatments are applied. After fermentation, a distillation is performed to increase the alcohol concentration in the biofuel.

1.3.2 Biofuel Production

The physicochemical processes allow the vegetable oils conversion into biodiesel (or methyl ester). These oils are found inside the seeds or fruits of energy crops; therefore, the first step of the process consists of the oil extraction, either by mechanical or chemical means. In the mechanical extraction, presses are used that allow to collect the oil, obtaining a "cake" rich in proteins and carbohydrates as a by-product. In chemical extraction, a solvent, usually hexane, is used to separate the oil from the rest of the components of the biomass. In this case, it is necessary to evaporate the solvent to obtain oil with a higher degree of purity. Chemical extraction allows to recover more oil than mechanical extraction and, for this reason, is the most used. After extraction, the oil is subjected to a transesterification process; whereby the triglycerides (formed by fatty acids) obtained from the oils react with an alcohol, in the presence of a basic or acid catalyst. The products obtained are glycerin and biodiesel with low degree of purity. After that, it is necessary to separate the biodiesel from the glycerin. This is done by decanting, since the glycerin is deposited at the bottom of the reactor. Subsequently, it is usual to wash the biodiesel with water, to eliminate some salts that can be formed during the reaction. Finally, the water is removed by decanting and/or evaporation and pure biodiesel is obtained.

1.4 Biofuels in Argentina

The development of biofuels in Argentina dates back to the end of '70s with the "Alconafta" Program, which was a pioneer project in the Latin American biofuels industry. The program was successful at the beginning, but it did not prosper for long due to high fiscal cost and political and economic troubles that the country was going through at that time. However, research and development has grown steadily since then in topics related to alternative production of biofuels intended for mixtures with fossil fuels for transportation. Research lines in the field of agro-industry, energy and environmental sciences were tackled as well. Since then, oilseeds production has been growing and this trend corresponds to a long-term path that accelerated in the last years.

As a result, Argentina became one of the world's largest biofuels producers and consequently, the largest exporter, especially of biodiesel (Timilsina et al. 2013). Currently, it is located number seven in the ranking of biofuel producing countries, and number four in the ranking of biodiesel producing countries. For this reason, energy from vegetable oil (in particular soybean biodiesel) is at the center of the attention of numerous social sectors: researchers and academics, politicians, industrialists, farmers and common people (Piastrellini et al. 2015). Regarding bioethanol, its development in Argentina is lower compared to the latter, but it is becoming a leading protagonist due to the blend mandates established since the entry in force of Argentinean law 26,093 in 2010. This law establishes that gasoil and gasoline marketed in the national territory must be mixed with biodiesel and bioethanol, respectively. Currently, the mandatory cut is 10% for diesel and 12% for gasoline. This means that diesel sold in Argentina must have 10% biodiesel (B10), while gasoline must contain 12% bioethanol (E12).

1.5 Environmental Problems Associated to Biofuels

The production and use of biofuels, specifically in the transportation sector, present many advantages compared to fossil fuels in regard to Climate Change and Greenhouse Gas emissions, which have supported the development of the former value chains. However, there are other environmental impacts associated to growing crops as feedstock for biofuels such as land use, water use, use of fertilizers and other agrochemicals, which could adversely impact both resources availability and quality (Dominguez-Faus et al. 2009). This is because the main raw materials destined to produce biofuels are also used (directly or indirectly) for the production of food.

This situation has generated a growing interest in the development of biofuels derived from non-food biomass and with low demand for resources, especially land and water. In this way, successive generations of biofuels have emerged: those of first generation derived from biomass with food uses and economically viable on an industrial scale (for example bioethanol from sugarcane or corn, and biodiesel from oil crops, as soybean and rapeseed); those of second generation derived from waste or non-food crops (such as lignocellulosic energy crops, mostly *Populus* spp., and perennial grasses, e.g., *Panicum virgatum*, *Miscanthus* spp., *Andropogon gerardii*, and *Arundo donax*); and the third generation produced from algae, mainly microalgae (Carneiro et al. 2017). Though there is an increasing interest in second generation biofuels, there are very few commercial industries and third generation biofuels are still at the lab stages.

This is one of the main reasons that justify a long list of studies that are intended for calculating the environmental impacts of biofuels (e.g. Amores et al. 2013; Daylan 2016; Gnansounou and Raman 2016; Harris et al. 2016; Morales et al. 2015), and paying special attention to the impacts on water use (Chiu et al. 2012; De Fraiture et al. 2008; Gerbens-Leenes and Hoekstra 2009; Nilsalab et al. 2017; Wu et al. 2014; Yang et al. 2011).

In general terms, first-generation biofuels are based on "hight-input" agricultural systems, that is, they require a large amount of agrochemicals, water and land to reach acceptable levels of productivity. This is the main cause of the environmental impacts associated with the production of these biofuels. Among second generation biofuels, perennial grasses stand as an interesting option because they have high biomass productivity, even in low quality sites; require minimal or no fertilizers input suitability, standard agricultural machinery can be used for their production; and energy balances are positive (Fernández 2003). According to Oliver et al. (2009), higher CO_2 atmospheric figures and severe drought conditions predicted by climate models developed by IPCC, would not affect their yield due to their C4 photosynthetic pathway.

1.6   The Use of Water in Biofuels Production

The increase in the use and consumption of freshwater on a global scale is related to the world population increase associated, in turn, with a greater production of agricultural products for domestic consumption. Likewise, many countries have increased the use of fresh water because they produce primary goods to export to other countries that do not have enough available land to produce them for themselves (Civit et al. 2011; Hoekstra and Chapagain 2006). The latter is precisely the case of Argentina, which exports water in a "virtual" way within agricultural products destined to the foreign market. The virtual water of a commodity is the volume of water that is used to produce it, extracted and measured in the place where it was produced. The other side of the coin is that many other countries prefer to import commodities that they could produce within their boundaries because they have decided not to compromise their freshwater reserves with other purposes than food production (Hoekstra and Chapagain 2006). It must be considered that, in most agricultural-based products, such as biofuels, the agricultural stage is usually the most relevant in terms of water use and consumption. Therefore, making a sustainable use of water resources, especially when producing energy from biomass sources, is essential since Argentina is a net exporter of virtual water, both in raw materials and agricultural-based products (Civit et al. 2011).

2   The Measure of Water Use in Biomass Sources for Biofuels. The Water Footprint

The concept of water footprint and its application in Latin America is relatively new (Martínez et al. 2016). At the beginning, the quantification of water footprint of agricultural-based products was mainly based on the methodology promoted by the Water Footprint Network (WFN), which quantifies the volumes of green, blue and grey water[1] associated to the production of a product, a service, a region, or even a person or group of people and its comparison with a scenario or reference value. Then, from 2015/2016 onwards, the application of the water footprint methodology developed by the International Organization for Standardization (ISO)—defined as "a metric that quantifies the potential environmental impacts related to water" (ISO 14046 2014)—has been widespread strongly, especially among the Life Cycle Assessment (LCA) community. This approach quantifies not only water volumes but also impacts on water availability (scarcity and degradation) and its consequences on human health, ecosystems and resources. It identifies the amount of water used and the changes occurred in water quality. Under the umbrella of the latter methodology, different results can be obtained depending on the objective of the study: it could be a stand-alone assessment or being part of a LCA. Always the term "water footprint" is

---

[1] According to the Hoekstra et al., blue water.

used when it is the result of an impact assessment. There are several methodologies that allow evaluating the impact of water use with an ISO approach (Pfister et al. 2009; Boulay et al. 2018).

To assess the scarcity water footprint of each biomass source considered in this chapter, the ISO Standards were adopted. ISO 14046 (2014) defines the principles, requirements and guidelines (ISO 14046 2014; Pfister et al. 2010).

3 Case Studies

The term oleaginous comes from the Latin *oleaginous* "oily" which is an adjective. This term is used to refer to plants that produce seeds or fruits from which oil (liquid triglycerides) is obtained. From the oleaginous, two biomass sources for obtaining biodiesel are considered: rapeseed (*Brassica napus*) and soybean (*Glycine max*). Producing bioethanol requires crops with high content of sugar, starch or lignocellulose. Within this group, two crops are selected: sugarcane (*Saccharum officinarum*) and a native grass (*Spartina argentinensis*). They are located in different ecoregions within Argentina, and therefore, with different edaphoclimatic features. Figure 1 shows the locations of each case study considered.

3.1 Functional Unit and System Boundaries

The functional unit (FU) chosen for all cases is 1 *MJ of biomass energy*. The system considers biomass production (agricultural stage) including soil preparation, sowing, irrigation (only for rapeseed), fertilization, management of crop pests, and harvesting (Fig. 2). All direct and indirect water uses, materials and fuels flows with a significant impact on outcome are included in the study.

As direct water use, irrigation water was considered (only for rapeseed since the other crops cover their water requirements with rainwater). As an indirect use of water, the water required for the production of fuels necessary for the tractor used in agricultural activities, of electricity and agrochemicals was taken into account. For the transportation of agrochemicals, two stages were considered. The first one stands for the way from the factory to the closest city to the plantation, by using 16 t trucks. The second stage runs about 30 km from the urban center to the cultivation area, carried out in small trucks. A generic fertilizer and a generic pesticide for each biomass are considered, with the aim of making comparable the four cases among them and with other results obtained in different studies. Water embodied in agricultural inputs was extracted from Ecoinvent Database (Wernet et al. 2016).

As this study represents a water availability footprint according to ISO 14046 (ISO 14046 2014), green and gray water consumption are excluded. A cut-off criteria of 0.01% was applied; therefore, the agrochemicals water dilutions are excluded from the system boundaries.

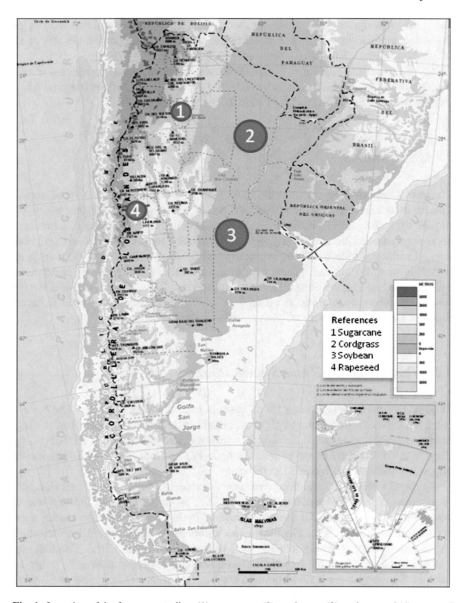

Fig. 1 Location of the four case studies: (1) sugarcane, (2) cordgrass, (3) soybean and (4) rapeseed

With the water inventory results, an impact assessment has been conducted in order to evaluate the regional consequences resulting from the fresh water extraction in the different considered regions. For doing so, the impact assessment model by

Fig. 2 System boundaries and value chain considered in the four case studies. *Source* Adapted from Lamers et al. (2008)

Pfister et al. (2009) has been applied, considering characterization factors at basin level (Pfister et al. 2010).

For all biomass sources considered, the water footprint estimation encompasses a one-year time window, using five-year average meteorological and yield data (2012–2016).

3.2 Sugarcane

The sugar world production of sugarcane during campaign 2017/2018 was around 190 million tons (USDA 2018a). Most sugar in the world (80%) is derived from sugarcane (*Saccharum officinarum*) and the remainder from sugar beet (*Beta vulgaris*). Sugarcane production takes place in tropical regions, particularly in developing countries in America, Asia, and Africa. Brazil and India are the first two sugarcane producers in the world. Argentina occupies position number 19, with a production rate close to Iran and South Africa. Regarding sugarcane-based ethanol, Brazil leads the worldwide production (Cremonez et al. 2015). The success of sugar cane as an energy crop lies in the fact that it is a species with high resistance, rapid growth, and high carbon uptake capacity. Therefore, the sugarcane-based ethanol might help to mitigate global warming. Nevertheless, there are some drawbacks associated with sugarcane intensification, such as land use change, competition with food, environmental impacts associated with the transportation sector, and the generation of large amounts of wastewater. This complex context has served to encourage studies leading to sustainable production of sugarcane.

Sugarcane is generally regarded as one of the most important and efficient sources of biomass for ethanol (biofuel) production. It offers productive alternatives to food (sugar), fiber (bagasse) and energy, particularly cogeneration of electricity and ethanol. Brazil is the major sugarcane-based fuel ethanol producing country, accounting for 28% of global demand (RFA 2017). Sugar mills in the world usually generate part of their energy needs from bagasse (the fibrous material remaining after sugarcane milling). Some of these facilities have high efficiency boilers which allow them to cogenerate electricity for their own needs and to sell to the local energy grid as well.

Argentina participates actively in the Paris Climate Agreement and COP22 against climate change. In support of this commitment, the country approved the National

Biofuels Law 26,093, active from 2010, which sets the percentage of biodiesel and bioethanol in diesel and gasoil, respectively. Moreover, a secondary objective of this measure is to diversify energy supply and to promote the development of rural zones with small- and medium-sized sugarcane growers. After several modifications along the years, the current mix mandated for bioethanol in gasoline is 12%. Bioethanol production for 2018 is expected to be 1120 million L. This production will be split in halves for sugarcane and corn. Nowadays, bioethanol production has reached an average figure of 11.7% in gasoline, just below the official mandate, always focusing the domestic market (USDA 2017a).

The Argentine sugar industry is based exclusively on sugarcane, being bioethanol its main co-product. Sugarcane production concentrates in the Northwest (Tucumán, Salta and Jujuy) and the Northeast. The sugarcane industry is formed by 23 sugar mills, 16 ethanol distilleries, and 9 plants for ethanol dehydration. This industry is one of the main economic activities of the province of Tucumán (15 sugar mills) (Argentine Sugar Center 2018), which hosts a plantation area for sugarcane of about 250,000 ha and a sugar production, during campaign 2017, of two thirds of the national production (USDA 2017b). There prevails a productive system formed by more than 6000 independent growers—many of them with less than 50 ha each—with different technology levels (TL) and different degree of access to the productive factors, which sell their production to the sugar factories by using different commercial schemes. This exchange structure between growers and sugar mills—a non-vertically integrated supply chain—is different from the main sugarcane areas of the world where sugarcane production and mills are vertically integrated (Acreche and Valeiro 2013).

It is noteworthy that sugarcane stands out by its socioeconomic relevance. In the sugarcane crop areas, it can be observed a higher degree of industrialization and an expansion of the productive infrastructure. This generates an increase of employment, especially during harvesting season. Additionally, the sugar sector promotes some other activities which develops around the agroindustry, such as trading and supply systems.

Sugarcane production in Tucumán started in the XVII century. Tucumán is the smallest Argentine province located in the north-west of the country. Sugarcane cultivated area is located on the west-center plains of the province (65.3°W, 27.0°S), in the vicinity of the Salí-Dulce River. The local climate is classified as subtropical with dry season in winter. The average annual temperature is 19 °C, with a maximum temperature, in summer, around 35 °C. The last one is the optimal temperature for the crop growing period. The driest month, July, usually has less than 9 mm of precipitation whereas the total annual rainfall is 1076 mm (Jorrat et al. 2018). Dominant soils in the crop area is silty-loam or silty in the surface (Romero et al. 2009). There are no reliable figures about irrigated surface in Tucumán and artificial irrigation can be considered as not significant.

In this study, three TLs have been considered for the agricultural labors: high (HTL), medium (MTL), and low (LTL), as defined by local researchers (Giancola et al. 2012). The approximate distribution of these TLs in the cultivated area is 40, 50, and 10%, respectively. The main features of each TL are shown in Table 1.

Table 1 Main features of the technology levels considered for the agricultural labors (Nishihara Hun et al. 2017)

Main differential aspects	High (HTL)	Medium (MTL)	Low (LTL)
Crop yield (t/ha)	75	62	55
Harvest system	Mechanized	Semi-mechanized	Semi-mechanized + manual
Trash burning	Scarce	Total	Total
Agrochemicals use	Intensive	Moderate	Scarce

The system description and inventory data are to a large extent the existing ones for sugarcane production in Argentina. System boundaries are expanded to include the impact associated with the production of all inputs (e.g., agrochemicals and fuel). Data regarding this activity are not fully available nor gathered into a unified database. Therefore, some data are based on average values. Information related to agricultural labors has been taken from local producers, the Argentine National Institute of Agricultural Technology (INTA), and other governmental institutions. Data gaps have been filled using specialized literature, handbooks, and databases (e.g., Ecoinvent v3.1). A portion of these data has already been used in other works by the authors (Amores et al. 2013; Nishihara Hun et al. 2017; Jorrat et al. 2018).

Most of the activities involved in sugarcane production such as planting, cultivation (ploughing, chiseling, furrowing, among others, corresponding to soil preparation, and pulverization), and harvesting, are included. Sugarcane (ready-to-transport sugarcane stalks leaving the crop fields) is regarded as the main product. Broadly, production of this crop is characterized by a system of hand plantation with low use of agrochemicals. Artificial irrigation is not significant (Romero et al. 2009). As a usual practice, sugarcane is allowed to grow with the same stalk several times after harvesting. The annual renewal percentage depends on the TL. The portion of the plantation formed by new plants is known as "cane plant", whereas the rest of the plantation, of two or more years old, is called "cane ratoon".

The pulverization tasks for cane plant consider the agrochemicals production process and fuel used for their application. In general, N fertilizer is supplied as urea and P fertilizer as triple super phosphate. No K fertilizer is applied in sugarcane production in Tucumán. Other agrochemicals used are ametryn, atrazine, and 2,4-D (Table 2). The type and dose of each agrochemical strongly depend on the TL. It has been considered that agrochemicals are produced in the country at a distance of 1500 km from Tucumán. Along the whole study, the production and distribution processes for diesel and electricity is adapted as much as possible to the Argentine context.

The process harvesting entails information mainly related to infrastructure and fuel used during harvesting, taking into account that the harvesting method also varies with the TL: mechanized (green harvest), semi-mechanized, or manual. In the first case, a harvester machine harvests, chops, cleans, and handles the cane in the form of billets, which are then loaded into a lorry. No pre-harvest burning is carried out. In the second method, sugarcane leaves and tops are burnt before harvesting to facilitate

Table 2 Agrochemicals consumption (basis: 1 ha harvested sugarcane)

	High (HTL)	Medium (MTL)	Low (LTL)
Urea (46% N) (kg)	216	188	178
Triple phosphate (45% P) (kg)	4.00	3.07	–
Atrazine (50% N) (L)	3.20	0.61	–
Ametryn (50% N) (kg)	1.00	1.27	–
Acetochlor (80% N) (L)	0.40	0.31	–
2,4-D (50% N) (L)	0.96	1.69	1.78
Monosodium methyl arsenate (69% N) (L)	0.80	1.02	–
Metolachor (90% N) (L)	0.80	–	–
Paraquat (90% N) (L)	0.16	–	–

the operation, then cane stalks are cut and a loader machine put the stalks into a lorry. The third method consists of manual felling, topping, de-trashing, bundling, and loading the stalks into the transportation vehicles. Usually, part of the remaining trash is burnt on the field (post-harvest burning). Emissions to air and water originates in the agrochemical transformation after being applied onto the crop: denitrification, volatilization, leaching, and runoff (Nemecek et al. 2007).

Sugarcane crop yield in Argentina depends on the TL (Table 1). The weighted crop yield for the region can be estimated in 67 t sugarcane per ha. Average yield of sugarcane in Brazil, the main reference for any sugarcane grower, between 2008 and 2014, is 75.2 t ha^{-1} (Global Yield Gap 2018), similar to the crop yield of Tucumán for the HTL.

Regarding representativeness of this case study in Argentina, it is important to highlight that being Tucumán the core of the sugarcane industry in the country, figures and discussion raised in this chapter are full representative of the current local situation (Fig. 3).

3.3   Cordgrass

Spartina argentinensis Parodi (=*Sporobolus spartinus* (Trin.) P. M. Peterson & Saarela) (cordgrass) is the main species of Argentinean herbaceous communities named "espartillares". These communities occupy large depressed areas (*circa* 4.5 million hectares), with halohydromorphic soils in the province of Santa Fe (28–30° S, 60–61° W, altitude: 21 m) (Lewis et al. 1990), plus flooding areas surrounding the Dulce River and the Laguna Mar Chiquita in the province of Córdoba (Oyarzabal et al. 2018). The espartillares are communities with a strong dominance of one or two species, which determines a low floristic diversity and a large number of species with very low relative importance. Lewis et al. (1990) reported that most of these communities have suffered less direct human disturbance than others in the region. The poor

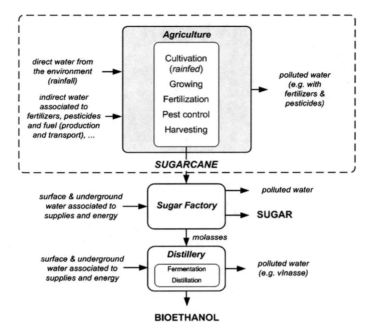

Fig. 3 Scheme of the system considered for sugarcane production

drainage of soils and its salt figures hinders agriculture in this region. Hence, raising cattle is the main productive activity within this region. Fire is a frequent practice carried out by farmers, since cattle prefer tender shoots rather than full grown leaves. Nevertheless, this practice does not only release CO_2 into the environment with no energy gain, but also affects the quality of life in the neighboring urban centers and with no substantial increase in cattle production, hence in economic activity in these areas.

Cordgrass is a rhizomatous perennial C4 grass in which leaf and culm emergence is promoted by both fire and clipping (Feldman and Lewis 2005, 2007), due to higher photosynthetic rates of leaves that continued growing after fire or clipping even under water deficit conditions. These high photosynthetic rate values remained high during two months after fire or clipping treatments enabling the production of high figures of tillers, leaves and biomass (Feldman et al. 2004).

A bioethanol yield of 241.43 L t^{-1} as reported by Jozami et al. (2013), considering 10 Mg ha^{-1} of annual net biomass production and 70% harvest efficiency is considered. Obtaining cordgrass biomass for bioethanol production from *S. argentinensis* involves the following field processes and inputs (see Fig. 4): (i) diesel for the processes of disking for soil smoothing (in order to turn down *Camponotus* sp. anthills, and grass bale production with a round baler); (ii) mowing, by rotary mower; (iii) bale production and loading, and (iv) transportation by truck (34 round bales of 700 kg per tuck, 30 km).

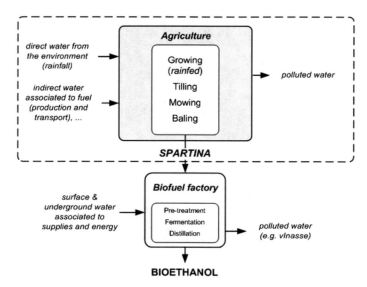

Fig. 4 Scheme of the system considered for cordgrass production

3.4 Soybean

The soybean world production during campaign 2017/2018 was around 334.81 mil-lion tons (USDA 2018b). United States was the leading soybean producing country with a production volume of 119.5 million tons, following by Brazil with 115 mil-lion tons. 2017/18 was one of the worst campaign for soybeans in Argentina with the majority of the summer crop experiencing drought conditions. The growing season ended with multiple, heavy rains that did not benefit yields and delayed harvest. However, Argentina ranks third in the ranking of soybean producing countries, with a volume produced of 40 million tons in the last campaign.

Biodiesel production is more geographically diverse than bioethanol, with pro-duction spread among many countries (REN 21 2017). In 2017, the leading countries for production of biodiesel were United States (18% of global production), Brazil (12%), and Indonesia, Germany and Argentina (each with 10%).

The biodiesel sector developed as a result of the world vegetable oil market (Souza et al. 2017). Currently, biodiesel is produced from well-established crops used in the food industry, such as rapeseed, soybean and sunflower. Yet, there are many other relevant candidates for biodiesel production able to add value by promoting the coproducts market, with further contribution to the diversification of the sector (Souza et al. 2017). In Argentina, biodiesel is produced exclusively from soybean oil. This situation is favored by the consolidation of the country soybean complex in the international market, granting the possibility of adding value and diversifying the export matrix. Approximately 56% of total production in the country is exported mainly to the United States and Europe, and the remaining 44% goes to the domestic

Table 3 National production and main destinations of soybean biodiesel

Time	Production	Sales to blend	Other sales to the domestic market	Exports
2008	711,864	0	265	680,219
2009	1,179,103	0	426	1,142,283
2010	1,820,385	503,325	5241	1,342,318
2011	2,429,964	739,486	9256	1,649,352
2012	2,456,578	824,394	50,400	1,543,094
2013	1,997,809	884,358	618	1,149,259
2014	2,584,290	969,456	685	1,602,695
2015	1,810,659	1,012,958	1403	788,226
2016	2,659,275	1,033,331	3069	1,626,264
2017	2,871,435	1,173,295	238	1,650,119

market, to cover the mandatory diesel blend or to generate electricity (Ministerio de Energía de la Nación 2017) (Table 3).

The entire Argentine soybean destined to produce biodiesel is cultivated in the Pampean region, situated in the central-eastern portion of the country. This region is home to the main vegetable oil and biodiesel hub of Argentina and has specific infrastructure for export through the Paraná-Uruguay waterway (Piastrellini et al. 2017).

In the Pampean region, 88% of the total cultivated area is under no-tillage (AAPRESID 2017). This agricultural technology does not harm the soil, often improving its physical, chemical and biological conditions, thus increasing productivity levels per hectare of occupied land. Around 70% of the area under no-tillage is sown between October and November (early soybean), and the remaining area during December (late soybean). Typically, the late soybean is planting after a winter crop and develops its cycle during a limited period, exposing itself to unfavorable environmental conditions (such as early frost, insufficient incident solar radiation or temperature). Therefore, crop yields are usually lower for late soybean than for early soybean. In addition, some production schemes respond to conventional tillage, which involves disking, plowing, and other methods of tilling up crop stubble left behind after harvest. This technology reduces the presence and incidence of pests, but increases the risk of soil erosion. The rainfall rate of the Pampean region allows soybean cultivation under rainfed conditions. However, there is an increase in the land area occupied by soybean under supplementary irrigation, usually supplied from groundwater sources (Piastrellini et al. 2015). Each of these agricultural schemes presents a substantial variation in soybean yields. The technical reports attribute yields close to 2800 kg ha^{-1} for early soybean in no-tillage; 2200 kg ha^{-1} for late soybean in no-tillage; 2380 kg ha^{-1} for early soybean in conventional tillage; and 3800 kg ha^{-1} when soybean is irrigated (Donato et al. 2008; Salinas et al. 2008; Martelloto 2012).

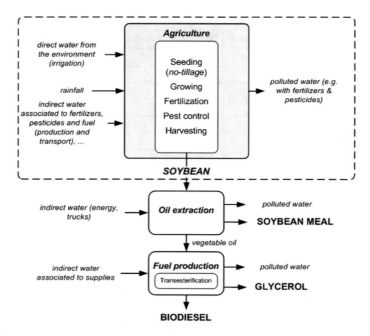

Fig. 5 Scheme of the system considered for soybean production

Particularly, the area under study is located in the northwest of Buenos Aires, in the center of the Pampean region. The system (Fig. 5) considers early soybean in no-tillage under rainfed conditions, and includes seed inoculation, sowing, fertilization, management of crop pests, and harvesting. Site preparation is not carried out, and stubble from the previous season is left on the soil surface. The water requirement of the soybean is completely supplied by the precipitation. The crop development cycle covers 120 days (initial stage, crop development stage, mid-season stage, and late season stage). The date of sowing is considered to be October 20th. The maximum radicle depth, in conditions of rainfed, is 1 m (Allen et al. 1998) and the crop mean height is 0.6 m. The residue cover corresponds to 80%. The rainfall rate of the study zone allows soybean cultivation under rainfed conditions; therefore it is not necessary to apply irrigation. Climatic data used correspond to the same period as the other biomass resources and are taken from the Junín-Buenos Aires weather station (34° 34′ S, 60° 95′ W, altitude: 82 m).

The average crop yield for this period is 2.28 t ha^{-1} (Ministerio de Agroindustria 2017). The Argiudol typic is the characteristic soil of the study region. The texture corresponds to a loamy clay soil. The field capacity and permanent wilting point are estimated with the SPAW Hydrology tool developed by USDA. The adjustment of Kc value due to the presence of stubble cover is carried out following the recommendations of Allen et al. (1998). This adjusted crop coefficient is needed to calculate the crop evapotranspiration considering the decrease of the soil moisture by evaporation. With the value of evapotranspiration, the real use of precipitation water is calculated.

It should be noted that this case study is fully representative of the recent situation in Argentina, since it is located in the soybean region par excellence of the country, and most of data considered to elaborate the life cycle inventory and calculate the environmental impact are local and current.

3.5 Rapeseed

The rapeseed (*Brassica napus*) is an oleaginous species belonging to the family of the Cruciferae, however, at present, from the taxonomic point of view, there are some authors who agree to include it within the family Brassicaceae.

This seed has been cultivated since ancient times. It is considered as one of the first crops used by the humankind. There is evidence of its consumption in India several centuries before the Christian era, passing to China and Japan at the beginning of our era. Later it was cultivated in Europe, due to its ability to grow and develop at low temperatures, which makes it one of the few oleaginous species suitable for cultivation in both warm-temperate and cold zones (Silva Colomer 2009).

This species was used primarily as industrial oil because of its good properties as a lubricant and also for lighting, since it produces a white flame without smoke. Its greatest demand came during the Second World War. Inedible rapeseed oil was used as a high temperature lubricant on steam boats, but with the shift to diesel engines in the following decade, industrial demand decreased. Farmers began looking for other uses for the plant and its products.

In the '50s, the demand as a food product of the rapeseed oil was insignificant due to some nutritional aspects of the oil, especially referred to its high erucic acid content. On the other hand, the obtained flour presented problems for the elaboration of food for animal consumption, mainly due to the presence of glucosinolates, responsible for the taste and smell characteristic of these plants, which produced nutritional problems in animal feed (Iriarte 2008).

The first cultivated rapeseed varieties contained a percentage of erucic acid in their oil, which ranged between 25 and 50%. But in the year 1966, Canadian scientists obtained the Gold variety with low content in this acid. The development of varieties with that characteristic constituted an important improvement in quality. This allowed the reduction to 2%, which is currently the amount of erucic acid contemplated in the world standard. By the year 1967, the genetic source of low glucosinolate content was discovered in a cultivar of Polish origin and incorporated into breeding programs.

In the early '70s, rapeseed varieties with low erucic acid content (LEAR) were developed. They also had a low glucosinolate content. The Western Canadian oilseed crusher association registered, under the name of "CANOLA" (Canadian Oil Low Acid), the varieties that possess such characteristics, to more easily identify this differentiated product. In 1973, Canadian agricultural scientists launched a marketing campaign to promote the consumption of canola (Thiyam-Holländer et al. 2013).

Today, rapeseed is grown for the production of animal feed, edible vegetable oils and biodiesel. The main producers include the European Union, Canada, China, India and Australia, producing 20.5, 19.6, 13.5, 7.1 and 4.1 million tons, respectively, during the season 2016–2017. According to the Department of Agriculture of the United States, rapeseed is the second oilseed produced worldwide (after soybeans) (European Commission 2017).

The last 20 years have witnessed a dramatic increase in the production of oilseeds worldwide, especially the production of soybeans, which has multiplied by a factor of 2.2 from 1992 to 2012. In the beginning of 2000, the production of rape seed surpassed that of cottonseed to become the second oilseed in the world (Carré and Pouzet, 2014). World trade of canola seeds and its products has also increased. The United States is the main importer of canola oil and flour because of its proximity to Canada and the ease of cross-border trade. Canada accounts for more than half of the world trade in canola, flour and oil seed. Canadian producers continue to expand the area and canola production. Recently, biodiesel producers in the EU have dramatically increased the demand for canola oil, and the processing capacity has expanded considerably. Although rapeseed production in the EU has expanded to support this industry, the EU has also become a major importer of rape seed (USDA 2018c). Also the demand outlook is strong due to the increasing use of vegetable oils in Japan, Mexico, China and India.

In Argentina, oil production is centered almost exclusively on summer crops. The rapeseed by its winter—spring cycle accesses the market at another time of the year, which supplies the industry at times when it remains idle and does not overlap with the milling of other oilseeds. As it is a crop of cold temperate areas, it offers the producer an important option as a component of its agricultural rotation, which in these areas is mainly limited to winter cereals. Rapeseed in direct seeding poses allows the realization of second crops such as soybeans or corn. The quality of the grain and the yield in oil that is obtained is of excellent quality which allows its acquisition by the most demanding markets. In the south and southeast center of the Province of Buenos Aires and east of La Pampa, winter and spring-type rape are produced. In this region, temperatures allow the cold conditions that the crop needs to be fulfilled to complete with all stages of development. The wheat regions coincide with the areas feasible for rapeseed cultivation. This area has a history in the cultivation of rapeseed and has a great potential, having reached its yields at 2800 kg/ha. However, in general, the average yields are of the order of 1300 to 1600 kg/ha (Iriarte and López, 2014; Iriarte 2008).

According to the Ministry of Agriculture in 2012, the national production of rapeseed has had great variations year after year. In 2010 a minimum of 12,700 ha were planted, and maximums of 92,700 ha in 2012, with a production of 23,000 and 128,000 t, respectively. Some of the reasons why the production does not reach a stable volume are the technological adjustments that must be developed year after year in the different environments, both from the genetic point of view, as integral management (planting, protection and harvest). The main destination of the rapeseed produced in Argentina is the European Union.

The province of Mendoza is located in the central western part of the country. It has a continental climate, sunny and dry throughout the territory, with relatively warm summers and cold winters. The soil, especially sandy loam, is arid due to scarce precipitation (between 150 and 350 mm per year). The productive activity is based, especially on the agriculture of perennial species in its oases, especially viticulture and fruit growing.

Mendoza has approximately 40% (approximately 178,732 ha) of its surface with systematized irrigation with certain possibilities to expand its crops, whether traditional or new crops (Silva Colomer et al. 2010). The rapeseed crop is presented as an activity that can be adapted to the climate and soil conditions of Mendoza and has demonstrated ample yield potential, both in grain and in oil, following a simple management scheme practiced in other regions of the country. Rapeseed is a winter/spring crop therefore it does not compete for irrigation water with most of the traditional crops that are spring summer; and in addition, it has little risk of being affected by hail and supports very well the icy characteristics of the area. To test the possibilities of cultivating rapeseed in the province, an experience was conducted whose results are used in this work. The study was carried out covering a complete production cycle to obtain rapeseed grown in an irrigated oasis in the department of Junín, province of Mendoza, Argentina (33° 7′ S, 68° 29′ W, altitude: 657 m). The focus is on the production of the seed and not on its storage, transportation and subsequent use. The crop development cycle took 193 days (initial stage, crop development stage, mid-season stage, late season stage). The date of sowing is considered to be in April while de harvest occurs in October. The maximum radicle depth was 0.20 m while the crop mean height was 1.5 m. The process analyzed was: land preparation, sowing, crop development, irrigation and harvest. The average seed production is 2.7 t/ha (Fig. 6).

During land preparation stage, pre-planting preparation tasks were carried out during the month of March. They included the weeding that was executed by means of a dredge driven by a tractor. Diammonium phosphate was added as a measure to counteract the nutritional deficiencies of the soil. The sowing of seed of the Legacy and Jura varieties was done manually, in rows of 15 cm in height and distanced 0.60 m each. In the stage of crop-development, the addition of the agrochemicals was considered (nitroguanimide, urea and dinitroaniline compounds). Their application was made manually. During the entire crop cycle, eight irrigations were carried out, distributed from the previous stage to the sowing until before the harvest. Surface irrigation water corresponding to the shift granted by the General Irrigation Department of the province of Mendoza. Irrigation was made by gravity; therefore no pump was used to water distribution. The crop plots were flooded until they reached approximately 150 mm in height and then the entry of water is stopped. The harvest of the pods was made towards the end of October.

A harvester tractioned by a tractor was used. The pods with the seeds were allowed to dry in the open air without the need for industrial dryers because the moisture content in the seeds usually is less than 3%, which presents no danger of mold formation. The dried seeds were stored for their later distribution in bulk.

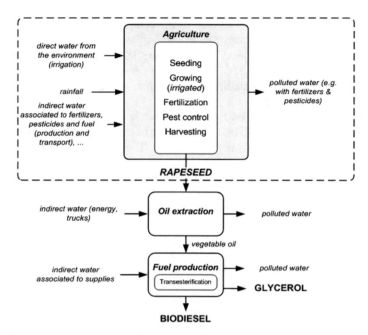

Fig. 6 Scheme of the system considered for rapeseed production

4 Water Scarcity Footprint Calculation

4.1 Inventory

The direct water use associated with the production of rapeseed is higher than figures obtained by the others sources of biomass analyzed in this study (Table 4). This is because rapeseed does not completely meet its water requirement with rainfall under Mendoza conditions. Therefore, it requires the addition of water by irrigation whereas the other crops do not, since they grow under rainfed conditions (Fig. 7).

With respect to the biomass types that can be used to produce biodiesel, the indirect water use associated with rapeseed production is approximately 400% higher than the one associated to soybean (Table 4 and Fig. 8). Although soybean demands a wide range of agrochemicals to achieve optimum crop yields, rapeseed also requires higher fertilizers and pesticides rates, which need more water for production. As for water incorporated in the fuel used in agricultural labors, it is greater for soybean than for rapeseed, in spite of soybean is produced in no-tillage, thus, not requiring soil preparation (Fig. 5). On the other hand, the water associated with the fuel production used to transport the inputs is greater for rapeseed due the longer distances traveled from the manufacture place to the agricultural site (1082 km for rapeseed versus 180 km for soybean) (Fig. 9).

Table 4 Water flows and water extraction for energy biomass production in Argentina

	Irrigated Rapeseed	Soybean	Sugarcane	Cordgrass
Direct flows (m^3/MJ):				
Precipitation	6.21E-04	9.12E-02	9.87E-03	9.40E-02
Irrigation	5.56E-01	0.00	0.00	0.00
Indirect flows (m^3/MJ):				
Supply chain	7.77E-02	1.54E-02	1.53E-03	0.00
Fuels	1.68E-04	2.89E-04	7.25E-05	1.72E-04
Transportation of inputs	2.41E-04	2.35E-05	1.59E-05	0.00
Total direct water use (m^3/MJ)	5.56E-01	0.00	0.00	0.00
Total indirect water use (m^3/MJ)	7.81E-02	1.57E-02	1.62E-03	1.72E-04
Total extraction (m^3/MJ)	6.34E-01	1.57E-02	1.62E-03	1.72E-04

Among the two biomass sources for ethanol production, the total water use for sugarcane production is approximately 845% greater than that of cordgrass (Table 4 and Fig. 10). This is because, unlike cordgrass, sugarcane requires fertilizers and pesticides to achieve optimum yields, which implies an additional water consumption to produce and transport agrochemicals. Although fuel use for cordgrass harvest doubles that of sugarcane, and hence it is more energy intensive in this aspect, the contribution of this process to the indirect water consumption is small (Fig. 11).

The direct water consumption values of soybean and sugarcane reported globally by Spang et al. (2014) are higher than those obtained in this study (Fig. 12). This is because the authors consider average production technologies from different countries, which include rainfed and irrigated systems; while in Argentina these crops are produced exclusively in rainfed areas. Conversely, Fig. 12 also shows that the water consumption value found in this study for rapeseed is around 28 times higher than the global estimates of Spang et al. (2014). This is explained because rapeseed considered in this study is produced in dry land, which implies the need for irrigation water; while Spang et al. (2014) take averages from different countries in which this crop is generally produced in wet land or requiring very little irrigation water.

Faist Emmenegger et al. (2011) reported values of direct water consumption for rapeseed produced in the arid region of Argentina lower than those found in this study (Fig. 12). Certainly, these authors have conducted the study taking into account data of crop water requirement based on edaphoclimatic conditions for season 2008/2009; while in this study the real value of water sheet applied (by irrigation) to the crop is considered.

Fig. 7 Isolines of precipitations in the four studied regions

4.2 Impact Assessment

The analysis of the biomass types used to produce biodiesel shows that, although water consumption (direct + indirect) of rapeseed is 40 times greater than soybean, results of the water scarcity assessment show that the impact value is approximately 400 times higher for rapeseed (Fig. 13).

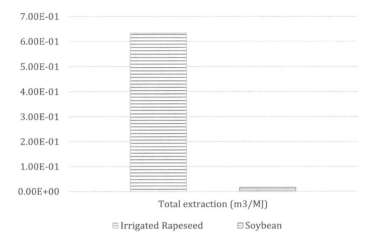

Fig. 8 Total water extraction for irrigated rapeseed and soybean production in Argentina

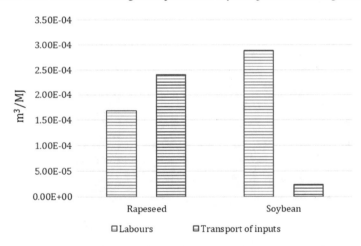

Fig. 9 Indirect water consumption associated with the fuel used in agricultural labour and transportation of agricultural inputs, for irrigated rapeseed and soybean production in Argentina

A similar situation is presented for biomass types that can be used for producing ethanol. The total water consumption of sugarcane is approximately 10 times greater than cordgrass, while the water scarcity impact value is 80 times greater (Fig. 14).

This is one of the particularities of the ISO approach, allowing us to consider the regional characteristics where each crop is produced. The main limitations of this study are related with the system boundaries on one hand, and with the degradation assessment, on the other. Results may vary significantly if the industrial stage is considered because in the case of cordgrass, for instance, where no irrigation and no water associated to agrochemicals, electricity or other field activities were taken

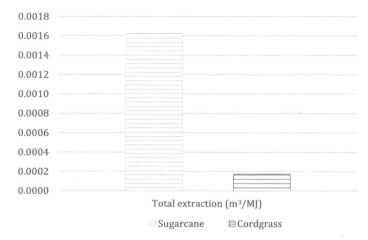

Fig. 10 Total water extraction for sugarcane and cordgrass production in Argentina

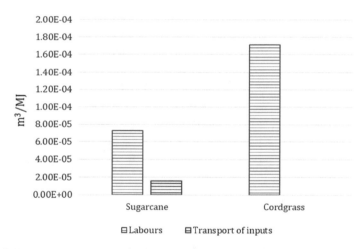

Fig. 11 Indirect water consumption associated with the fuel used in agricultural labor and transportation of agricultural inputs, for sugarcane and cordgrass production in Argentina

into account in the inventory phase. But in the industrial operations required to dry, cut, and transform the grass into ethanol and heat or syngas, consumptions may appear that increase the water footprint to values unknown because until now, the complete process has not been carried out neither in Argentina on an industrial scale elsewhere. There are only some experimental data and some other estimated data from the literature, which have not yet been validated in practice (Jozami et al. 2013). The other relevant limitation is that this study does not show the complete water footprint profile of biomass sources to obtain bioenergy because it does not

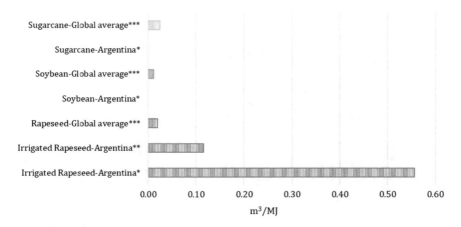

* Data found in this study.
** Data reported by Faist Emmenger *et al.* (2011).
*** Data reported by Spang *et al.* (2014).

Fig. 12 Direct water consumption for different sources of biomass

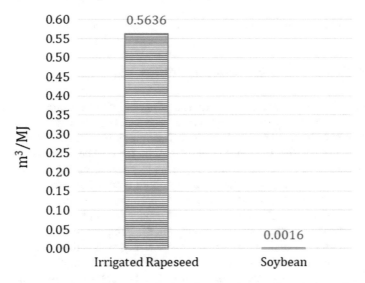

Fig. 13 Water scarcity impact for the production of rapeseed and soybean in Argentina

assess the degradation component. An important "next step" would be to consider the degradative conditions of the four case studies with the aim at obtaining a full water footprint profile as is suggested by ISO 14046 (ISO 14046 2014).

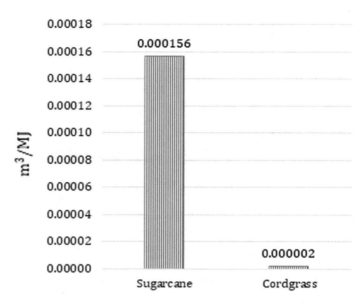

Fig. 14 Water scarcity impact for the production of sugarcane and cordgrass in Argentina

5   Conclusions and Recommendations

Scarce information is available on water footprint of biofuels produced in Argentina. A comparison of the water consumption and associated impacts of the production of 1 MJ of energy biomass is considered in different regions of the country.

The results show that the types of biomass with the best performance in terms of water consumption in Argentina are: soybeans to produce biodiesel and cordgrass to produce ethanol. The good performance of both types of biomass in comparison with the substitutes is the result of the low requirement of inputs and labors and the optimal edaphoclimatic characteristics of the production regions.

Both soybean and cordgrass are produced in the Pampean Region—a highly productive food region—and Chaco Region, respectively, so it is recommended to optimize the management of water resources in these regions to minimize competition with food products. Anyway, as *Spartina* is a native grass that grows naturally in the mentioned region, does not compete with food or livestock feed, because after removal for bioenergy, cordgrass would continue growing, producing tender eaves which could be used for cattle rising. An aspect that still needs to be established, is how long should be periods between one removal for bioenergy from the other in order to secure that the communities remain resilient (Feldman and Lewis 2005).

On the other hand, considering the environmental fragility, it is recommended to avoid the production of biomass destined to bioenergy in arid zones of the country.

It is worth noting that, for the sake of sustainability, there is a compelling need to complete the assessment of the entire environmental footprint of the biofuels (both the agricultural and industrial stages), being the water footprint an important indicator involved.

To acquire a deeper insight on sustainability aspects, the impacts related to climate change, biodiversity loss, land use, resources depletion, etc., cannot be left aside.

Considering what Spang et al. (2014) state in their article "given the lack of a comprehensive international database of biofuel production by crop feedstock, we needed to derive estimates for biofuel feedstock production for each country based on secondary reports", we think that this study provides a solid basis for conducting comparative water footprint studies between the biofuels produced in Argentina including local information and in most cases, primary data. On the other hand, this study opens the window to further researches that allow to have new insights on water degradation impacts, and actualized figures to compare biofuels and the fossil fuels they replace. Finally, the set of results obtained could become a strong tool for decision makers in the field of energy and energy management.

Acknowledgements This study was addressed by three research groups belonging to National Universities and the National Council of Scientific and Technical Research, in different regions of the country. Therefore, each one addressed a case of biomass source according to the conditions and aptitudes of the region considered.

References

AAPRESID. (2017). Evolución de la superficie en siembra directa en Argentina. http://www.aapres id.org.ar/wp-content/uploads/2013/02/aapresid.evolucionsuperficie_sd_argentina.pdf. Accessed March 22, 2018.

Acreche, M. M., & Valeiro, A. H. (2013). Greenhouse gasses emissions and energy balances of a non-vertically integrated sugar and ethanol supply chain: a case study in Argentina. *Energy, 54,* 146–154.

Allen, R. G., Pereira, L. S., Raes, D., & Smith, M. (1998). *Crop evapotranspiration: Guidelines for computing crop water requirements*. Irrigation and Drainage Paper No. 56. Rome, Italy: FAO.

Amores, M. J., Mele, F. D., Jiménez, L., & Castells, F. (2013). Life cycle assessment of fuel ethanol from sugarcane in Argentina. *International Journal of Life Cycle Assessment, 18*(7), 1344–1357.

Argentine Sugar Center. (2018). Available at www.centroazucarero.com.ar. Accessed May 1, 2018.

Boulay, A. M., Bare, J., Benini, L., Berger, M., Lathuillière, M. J., Manzardo, A., et al. (2018). The WULCA consensus characterization model for water scarcity footprints: assessing impacts of water consumption based on available water remaining (AWARE). *International Journal of Life Cycle Assessment, 23*(2), 368–378. https://doi.org/10.1007/s11367-017-1333-8.

Carré, Patrick, & Pouzet, André. (2014). Rapeseed market, worldwide and in Europe. *OCL, 21*(1), D102. https://doi.org/10.1051/ocl/2013054.

Carneiro, M. L. N., Pradelle, F., Braga, S. L., Gomes, M. S. P., Martins, A. R. F., Turkovics, F., et al. (2017). Potential of biofuels from algae: Comparison with fossil fuels, ethanol and biodiesel in Europe and Brazil through life cycle assessment (LCA). *Renewable and Sustainable Energy Reviews, 73,* 632–653.

Chiu, Y.-W., Suh, S., Pfister, S., Hellweg, S., & Koehler, A. (2012). Measuring ecological impact of water consumption by bioethanol using life cycle impact assessment. *The International Journal of Life Cycle Assessment, 17*(1), 16–24. https://doi.org/10.1007/s11367-011-0328-0.

Civit, B., Arena, A. P., Piastrellini, R., Curadelli, S., & Silva Colomer, J. (2011). Comparación entre la Huella Hídrica de biodiesel obtenido a partir de aceite de colza y aceite de soja. Revista de la Asociación Argentina de Energías Renovables, 15. Printed in Argentina. ISSN 0329-5184.

Cremonez, P. A., Feroldi, M., Feiden, A., Teleken, J. G., Gris, D. J., Dieter, J., et al. (2015). Current scenario and prospects of use of liquid biofuels in South America. *Renewable and Sustainable Energy Reviews, 43,* 352–362. https://doi.org/10.1016/j.rser.2014.11.064.

Daylan, B. (2016). Life cycle assessment and environmental life cycle costing analysis of lignocellulosic bioethanol as an alternative transportation fuel. *Renewable Energy, 42,* 1349–1361. https://www.sciencedirect.com/science/article/pii/S0960148115304754.

De Fraiture, C., Giordano, M., & Liao, Y. (2008). Biofuels and implications for agricultural water use: Blue impacts of green energy. *Water Policy, 10*(1), 67–81. https://doi.org/10.2166/wp.200 8.054.

Dominguez-Faus, R., Powers, S., & Burken, J. (2009). The water footprint of biofuels: A drink or drive issue? *Environmental Science and Technology, 43,* 3005–3010.

Donato, L., Huerga, I., & Hilbert, A. 2008. Balance Energético de la Producción de Biodiesel a Partir de Aceite de Soja en la República Argentina, INTA Report No. IIR-BC-INF-08-08. Buenos Aires, Argentina.

European Comission 2017. Oilseeds and protein crops: Market situation. In *Committee for the common organisation of agricultural markets*. October 25, 2017.

Faist Emmenegger, M., Pfister, S., Koehler, A., de Giovanetti, L., Arena, A. P., & Zah, R. (2011). Taking into account water use impacts in the LCA of biofuels: An Argentinean case study. *The International Journal of Life Cycle Assessment, 16*(9), 869–877.

Feldman, S. R., Bisaro, V., & Lewis, J. P. (2004). Photosynthetic and growth responses to fire of the subtropical-temperate grass *Spartina argentinensis* Parodi. *Flora, 199*(6), 491–499.

Feldman, S. R., & Lewis, J. P. (2005). Effect of fire on the structure and diversity of a *Spartina argentinensis* tall grassland. *Applied Vegetation Science, 8,* 77–84.

Feldman, S. R., & Lewis, J. P. (2007). Effect of fire on *Spartina argentinensis* Parodi demographic characteristics. *Wetlands, 27*(4), 785–793.

Fernández, J. (2003). Energía de la biomasa. In J. M. de Juana (Ed.), *Energías renovables para el desarrollo*. Spain: Thompson-Paraninfo.

Gerbens-Leenes, P., & Hoekstra, A. (2009). The water footprint of energy from biomass: A quantitative assessment and consequences of an increasing share of bio-energy in energy supply. Ecological Economics. Retrieved from https://www.sciencedirect.com/science/article/pii/S0921 80090800339X.

Giancola, S. I., Morandi, J. L., Gatti, N., Di Giano, S., Dowbley, V., & Biaggi, C. (2012). *Causas que afectan la adopción de tecnología en pequeños y medianos productores de caña de azúcar de la Provincia de Tucumán. Enfoque cualitativo*. Buenos Aires: Ediciones INTA.

Global Yield Gap Data. (2018) http://www.yieldgap.org/brazil. Retrieved April 23, 2018.

Gnansounou, E., & Raman, J. (2016). Life cycle assessment of algae biodiesel and its co-products. *Applied Energy, 161,* 300–308.

Harris, T. M., Hottle, T. A., Soratana, K., Klane, J., & Landis, A. (2016). Life cycle assessment of sunflower cultivation on abandoned mine land for biodiesel production. *Journal of Cleaner Production, 112*(1), 182–195. Retrieved from https://www.sciencedirect.com/science/article/pii/S0959652615012846.

Hoekstra, A., & Chapagain, A. (2006). Water footprints of nations: Water use by people as a function of their consumption pattern. *Water Resources Management*. https://doi.org/10.1007/s11269-00 6-9039-x.

Iriarte, L. (2008). Cultivo de Colza. Editores: Ing. Agr. Liliana Iriarte/Ing. Agr. Omar Valetti. 1ª ed. Buenos Aires: Tres Arroyos. INTA publicaciones Nacionales, Chacra Experimental Integrada Barrow.

Iriarte, L., & López, Z. (2014). El cultivo de colza en Argentina. Situación actual y perspectivas. Chacra Experimental Integrada Barrow (Convenio MAA-INTA). 11 p. [en línea] http://inta.gob.ar/documentos/el-cultivo-de-colza-en-argentina-situación-actual-y-perspectivas.

ISO 14046. (2014). Environmental management—Water footprint—Principles, requirements and guidelines.

Jorrat, M. M., Araujo, P. Z., & Mele, F. D. (2018). Sugarcane water footprint in the province of Tucumán, Argentina. Comparison between different management practices. *Journal of Cleaner Production, 188,* 521–529.

Jozami, E., Sosa, L. L., & Feldman, S. R. (2013). *Spartina argentinensis* as feedstock for bioetanol. *Applied Technologies and Innovations, 9*(2), 37–44.

Lamers, P., McCormick, K., & Hilbert, J. (2008). The emerging liquid biofuel market in Argentina: Implications for domestic demand and international trade. *Energy Policy, 36,* 1479–1490.

Lewis, J. P., Pire, E. F., Prado, D. E., Stofella, S. L., Franceschi, E. A., & Carnevale, N. J. (1990). Plant communities and phytogeographical position of a large depression in the Great Chaco, Argentina. *Vegetatio, 86,* 25–38.

Martelloto, E. (2012). Potencialidad y limitantes del riego complementario. Segundo Seminario, Recursos Hídricos para el sector Rural, Argentina. September 11, 2012.

Martínez, A., Chargoy, J., Puerto, M., Suppen, N., Rojas, D., Alfaro, S. et al. (2016). Huella de Agua (ISO 14046) en América Latina, análisis y recomendaciones para una coherencia regional. Centro de Análisis de Ciclo de Vida y Diseño Sustentable CADIS, Embajada de Suiza en Colombia, Agencia Suiza para la Cooperación y el Desarrollo COSUDE. 90.

Ministerio de Energía y Minería-Presidencia de la Nación Argentina. (2017). Producción, ventas al mercado interno y exportaciones de biocombustibles. https://datos.minem.gob.ar/dataset/estadisticas-de-biodiesel-y-bioetanol. Accessed February 25, 2018.

Morales, M., Quintero, J., Conejeros, R., & Aroca, G. (2015). Life cycle assessment of lignocellulosic bioethanol: Environmental impacts and energy balance. *Renewable and Sustainable Energy Reviews, 42,* 1349–1361.

Nemecek, T., Heil, A., Huguenin, O., Meier, S., Erzinger, S., Blaser, S., et al. (2007). Life cycle inventories of agricultural production systems. Ecoinvent report No. 15, v2.0. Agroscope FAL Reckenholz and FAT Taenikon, Swiss Centre for Life Cycle Inventories, Dübendorf.

Nilsalab, P., Gheewala, S. H., Mungkung, R., Perret, S. R., Silalertruksa, T., & Bonnet, S. (2017). Water demand and stress from oil palm-based biodiesel production in Thailand. *The International Journal of Life Cycle Assessment, 22*(11), 1666–1677. https://doi.org/10.1007/s11367-016-1213-7.

Nishihara Hun, A. L., Mele, F. D., & Pérez, G. A. (2017). A comparative life cycle assessment of the sugarcane value chain in the province of Tucumán (Argentina) considering different technology levels. *International Journal of Life Cycle Assessment, 22*(4), 502–515. https://doi.org/10.1007/s11367-016-1047-3.

Oliver, R. J., Finch, N. W., & Taylor, G. (2009). Second generation bioenergy crops and climate change: a review of the effects of elevated atmospheric CO_2 and drought on water use and implications for yield. *GCB Bioenergy, 1,* 97–114.

Oyarzabal, M., Clavijo, J., Oakley, L. J., Biganzoli, F., Tognetti, P., Barberis, I. M., et al. (2018). Unidades de Vegetación de Argentina. *Ecología Austral, 28,* 40–63.

Pfister, S., Curran, M., Koehler, A., & Hellweg Pfister, S. (2010). Trade-offs between land and water use: regionalized impacts of energy crops. https://www.ethz.ch/content/dam/ethz/special-interest/baug/ifu/eco-systems-design-dam/documents/downloads/ei99/ifu-esd-EI99-LCAfood2010_pfister.pdf.

Pfister, S., Koehler, A., & Hellweg, S. (2009). Assessing the environmental impacts of freshwater consumption in LCA. *Environmental Science and Technology, 43*(11), 4098–4104. https://doi.org/10.1021/es802423e.

Piastrellini, R. (2015). Aportes a la determinación de la huella ambiental de biocombustibles en Argentina. Influencia de los sistemas de manejo de cultivos sobre el impacto del consumo de agua, del uso del suelo y de las emisiones de gases de efecto invernadero para el biodiesel de soja. PhD Thesis. Universidad Tecnológica Nacional, Mendoza, Argentina.

Piastrellini, R., Arena, A. P., & Civit, B. (2017). Energy life-cycle analysis of soybean biodiesel: Effects of tillage and water management. *Energy, 126,* 13–20.

Piastrellini, R., Civit, B., & Arena, A. P. (2015). Influence of Agricultural practices on biotic production potential and climate regulation potential. A case study for life cycle assessment of soybean (*Glycine max*) in Argentina. *Sustainability, 7,* 4386–4410. https://doi.org/10.3390/su70 44386.

REN 21. (2017). *Renewables 2017*. Global Status Report. http://www.ren21.net/wp-content/uploa ds/2017/06/17-8399_GSR_2017_Full_Report_0621_Opt.pdf. Accessed February 25, 2018.

RFA, Renewable Fuels Association (2017). *Industry statistics*. Available at www.ethanolrfa.org/re sources/industry/statistics/#145409899-6479-8715d404-e546. Retrieved April 23, 2017.

Romero, E. R., Digonzelli, P. A., & Scandaliaris, J. (2009). *Manual del Cañero*. San Miguel de Tucumán: Estación Experimental Agroindustrial Obispo Colombres.

Salinas, A., Martellotto, E., Giubergia, J. P., Álvarez, C., & Lovera, E. (2008). Soja: Evaluación de Cultivares con Riego Suplementario. http://www.elsitioagricola.com/articulos/salinas. Accessed March 12, 2018.

Sánchez Godoy, F. (2012). Potencial del cultivo de la chumbera (Opuntia ficus-indica (L) Miller) para la obtención de biocombustibles. PhD Thesis. ETSI Agrónomos, UPM. Madrid, España.

Silva Colomer, J. (2009). INFORME ANUAL. Proyecto Integrado "Apoyo al desarrollo sustentable de las empresas familiares y Pymes agropecuarias del Noreste de la provincia de Mendoza, basado en la diversificación productiva y el asociativismo" EEA Junin. Instituto Nacional de Tecnología Agraria (INTA).

Silva Colomer J., Castillo, J., Iriarte, L., & Villegas, N. (2010). Cultivo de colza bajo riego en Mendoza. INTA. https://inta.gob.ar/sites/.../script-tmp-cultivo_de_colza_bajo_riego_en_mendo za.pdf.

Souza, S. P., Seabra, J. E., & Nogueira, L. A. H. (2017). Feedstocks for biodiesel production: Brazilian and global perspectives. *Biofuels*, 1–24.

Spang, E. S., Moomaw, W. R., Gallagher, K. S., Kirshen, P. H., & Marks, D. H. (2014). The water consumption of energy production: an international comparison. *Environmental Research Letters, 9*(10), 1–14.

Thiyam-Holländer, U., Eskin, M., & Matthäus, B. (2013). *Canola and rapeseed: Production, processing, food quality, and nutrition* (p. 4). Boca Raton, FL: CRC Press. ISBN 9781466513884. Retrieved November 25, 2015.

Timilsina, G., Chisari, O., & Romero, C. (2013). Economy-wide impacts of biofuels in Argentina. *Energy Policy, 55,* 636–647. https://doi.org/10.1016/j.enpol.2012.12.060.

USDA. (2017a). *GAIN report: Argentina biofuels annual*. United States Department of Agriculture, Foreign Agricultural Service. July 2017.

USDA. (2017b). *GAIN report: Argentina sugar annual*. United States Department of Agriculture, Foreign Agricultural Service. April 2017.

USDA. (2018a). Sugar: World markets and trade. United States Department of Agriculture, Foreign Agricultural Service, May 2018.

USDA. (2018b). Soybeans. data & analysis. Available in https://www.fas.usda.gov/commodities/s oybeans. Accesed June 15, 2018.

USDA. (2018c). *Department of agriculture*. Economic Research Service. Available in https://www. ers.usda.gov/topics/crops/soybeans-oil-crops/canola.aspx. Accessed June 5, 2018.

Wernet, G., Bauer, C., Steubing, B., Reinhard, J., Moreno-Ruiz, E., & Weidema, B. (2016). The ecoinvent database version 3 (part I): Overview and methodology. *The International Journal of Life Cycle Assessment, 21*(9), 1218–1230. Retrieved from http://link.springer.com/10.1007/s113 67-016-1087-8.

Wu, M., Zhang, Z., & Chiu, Y. (2014). Life-cycle water quantity and water quality implications of biofuels. *Current Sustainable/Renewable Energy Reports, 1*(1), 3–10. https://doi.org/10.1007/s4 0518-013-0001-2.

Yang, J., Xu, M., Zhang, X., Hu, Q., Sommerfeld, M., & Chen, Y. (2011). Life-cycle analysis on biodiesel production from microalgae: Water footprint and nutrients balance. *Bioresource Technology, 102*(1), 159–165. https://doi.org/10.1016/j.biortech.2010.07.017.

Water Footprints of Hydropower Projects

Himanshu Nautiyal, Varun Goel and Paramvir Singh

Abstract As the world is shifting towards renewable energy power generation it becomes important to explore the correct picture of renewable energy technologies. Today every nation is ramping up investments in power sector and trying to increase its installed capacity of power generation through renewable energy due to benefits like less greenhouse gas emissions, less environmental impacts etc. associated with it. But at the same time it is being observed that it is important to assess the renewable energy technologies to find out their real environmental impacts. Among all renewable energy sources, hydropower is becoming the most popular and promising source throughout the world. The power generation through hydropower is solely depends on the availability of water resources so formation of large dams and reservoirs are associated with establishment of a hydropower plant. The effects of climate change affect the water resources throughout the world as well. Large scale reservoirs can be found as key players for low carbon and sustainable society in future. Consequently, it becomes important to assess the consumption of water from hydropower generation in terms of water footprints. This chapter reviews the estimation of water footprints of hydropower plants through different methodologies and come out to the conclusion that there is still requirement of an effective methodology for water footprints calculation of hydropower projects.

Keywords Water · Renewable · Footprints · Hydropower · Energy · Reservoirs Power · Dams

H. Nautiyal (✉)
THDC Institute of Hydropower Engineering and Technology, Tehri, India
e-mail: h2nautiyal@gmail.com

V. Goel · P. Singh
National Institute of Technology Hamirpur, Hamirpur, Himachal Pradesh, India
e-mail: varun7go@gmail.com

P. Singh
e-mail: param016@gmail.com

© Springer Nature Singapore Pte Ltd. 2019
S. S. Muthu (ed.), *Environmental Water Footprints*, Environmental Footprints and Eco-design of Products and Processes,
https://doi.org/10.1007/978-981-13-2739-1_2

1 Introduction

Water is the basic ingredient in all areas of human, domestic and industrial activities from a needle to a large ship. Energy sector is not untouched from the huge requirement of water. As water crises and degradation of ecosystem are becoming a big problem in many nations of the world so it becomes important to look up and analyse the current power generation technologies and activities to explore some solutions of water stress in further development in the power sector. Even a balance between power generation (energy) and required water consumption is essentially needed for controlling water scarcity in future. Hydropower is always being considered as one of the most promising sources of sustainable power generation and many nations in the world are trying to increase their power generation capacity through hydropower projects. Hydropower is generated by converting the energy of flowing water into electricity and used in variety of domestic and commercial applications. It is the one of largest source of renewable energy based power generation throughout the world. The power generation through hydropower projects accounts for about 16.6% of the global renewable energy power generation (Renewable 2017, Global Status Report). Table 1 shows a list of top nine countries in hydropower capacity addition in 2016.

There are several benefits associated with hydropower projects however some negative aspects are also associated with them as shown in Fig. 1. Hydroelectric power generation covers large scale hydropower dams and small run-of river types hydropower plants. However, environmental issues are associated with both of them. Land requirement for building large reservoir is one of the major issues in deployment of hydropower projects. In fact the volume of the reservoir may depend upon is installation capacity and land topography. Construction of dams in plane areas consume more land areas while benefit of getting reservoirs of greater depth in smaller areas can be achieved in hilly regions. Further the land submerged in the reservoir contributes in destruction of forest land, wildlife, arable land and rehabilitation of

Table 1 Top nine countries in hydropower capacity addition (Renewable 2017, Global Status Report)

S. No.	Country	Capacity added, 2016 (GW)
1	China	8.9
2	Brazil	5.3
3	Ecuador	2.0
4	Ethiopia	1.5
5	Vietnam	1.1
6	Peru	1.0
7	Turkey	0.8
8	Lao PDR	0.7
9	Malaysia	0.6

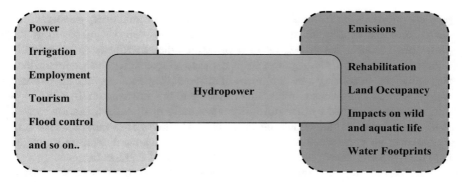

Fig. 1 Hydropower: benefits and impediments

communities in large scale. The construction of large dams has notable impacts on wild life and aquatic life too. There are always possibilities of destruction of aquatic organisms in power plant machineries; and interruption of flow due to large water storage reduces the flow at downstream and affects the aquatic life of the region. This is due to the fact that most of the hydropower plants are construed across the rivers and there are always great chances of entering aquatic organisms into the turbines. In addition to this, huge amount of water stored in the reservoirs promotes formation of algae blooms and weeds due to the deposition of sediments and other nutrients in large scale and cause imbalance in aquatic life.

Previous studies suggested high amount of GHG emissions associated with construction, installation and operation of hydropower projects. These emissions depend on various factors however capacity and size of the power plant; and type of submerged land are quite significant. The decomposition of vegetation and soil in the flooded area is responsible for producing methane, CO_2 etc. which further contribute in the GHG emissions associated with hydropower projects (Varun et al. 2009, 2010, 2012).

Power generation from hydropower is associated with large reservoirs and is solely dependent on availability of water. Scarcity of water in a hydropower plant always leads to reduction in power generation. However, dams associated with hydropower projects may be classified as single purpose and multi-purpose. Single purpose dams are those which are solely used for a specific purpose e.g. power generation only; whereas multi-purpose dams are used for more than one purpose like power generation and irrigation etc. One of the main concerns of the today is climate change problem which may affect earth's water resources in future. Therefore it becomes important to review the further growth of hydropower projects in respect to their water consumption.

Construction of big dams, occupancy of land, and flooded vegetation like impediments associated with hydropower projects have always been a big topic of debate related to its sustainable nature among environmentalist and engineers (Gleick 1992). Involvement of huge amount of water accumulated in large dams of hydropower projects questions its sustainable nature over other renewable energy sources like wind, solar etc. In addition to this, the construction of dams in upstream of the power plant reduces the flow to downstream which directly affects the aquatic life of downstream region. Hydropower dams exposed a large water surface area which is responsible for causing a large amount of artificial evaporation to the atmosphere and thus added water consumption rather flowing to the downstream. In addition to this, water is consumed by the land surfaces naturally by direct evaporation as well as transpiration from indigenous vegetation (combine known as evapotranspiration), (Lampert et al. greet.es.anl.gov) so construction of dams submerge the vegetated land nearby the reservoir and affect evapotranspiration of that location.

After considering all above points it becomes important to evaluate water footprints of the existing hydropower projects to figure out their correct picture as sustainable power generation source in future. Also, the importance of water footprints evaluation of hydropower lies on finding out environmental impacts of hydropower dams on aquatic life in downstream side and other water end users. High values of water footprints in hydropower sector may begin reputation risk for this sector which can further cause other investment risks in hydropower development in future (Bakken et al. 2013). Therefore, the main objective of water footprint assessment is to check the water perspective sustainability of hydropower facilities. The current chapter presents an overview of water footprints associated with hydropower projects. The discussion begins with the introduction and necessity of estimating water footprints of different hydropower facilities. Past studies pertaining to water footprints of hydropower projects are discussed which is then followed by discussions and conclusion.

2   Water Footprints

Water footprints may be defined as the amount of water required to produce a good or service. The estimation of water footprints comprises of direct and indirect water consumption required by a product, process etc. throughout the complete life cycle from the supply chain to the consumer. The concept of water footprint was first introduced by Hoekstra in 2002 which may be defined as an indication of freshwater consumption (both direct and indirect) by a consumer or producer (Hoekstra 2003). Water footprint for a product is expressed as the amount (volume) of fresh water required by a product throughout its full supply chain. The water footprints associated with a product are generally treated as a local or regional impact rather than a global impact like carbon footprints. This is because the consumption of water associated with a product is mainly fulfilled by the nearby water supply or resources.

The consumption of water is estimated in volume of water consumed (or evaporated) and or polluted per unit time (Hoekstra and Chapagain 2007). The water footprints can be used to estimate the amount of water consumed by an individual or whole community. It also incorporates the direct water consumption means the amount of water consumed directly by an individual; and indirect water consumption which is the total aggregate of the amount of water consumed by all the products (Fig. 2) (www.waterfootprint.org).

In general, water footprints can be classified as green water footprints, blue water footprints and grey water footprints. Blue water footprints can be defined as the volume of water origin from surface and groundwater resources like rivers, lakes, wetlands etc. and is either evaporated and used into a product. Green water footprints are estimated as the quantity of water from precipitation which has been absorbed and stored in root zone of soil and is either lost by evapotranspiration or utilised by the plants. Grey water footprints are defined as the volume of water consumed to manage and dilute the human and industrial discharges into the environment (www.waterfo otprint.org). The estimation of water footprints for a process, users, community and geographic location is expressed as volume of water per unit time (volumetric flow rate). Water footprints for a product are carried out by dividing the total summation of water footprints of different processes in its complete supply chain by the number of products (Water footprint assessment Manual 2011).

The concept of water footprint provides an effective view point for the relationship between a producer or consumer and the freshwater consumption. It helps to estimate quantitative figures of water consumption and pollution; and thus help to analyse their environmental impacts. However, the local impacts associated with consumption of water and pollution may vary with permeability of the water system and its consumers and other end users. The main benefit of water footprint estimation is to provide an ostensible analysis regarding the appropriate use of water for different domestic and industrial activities of human beings. This further helps to

Fig. 2 Water footprints assessment

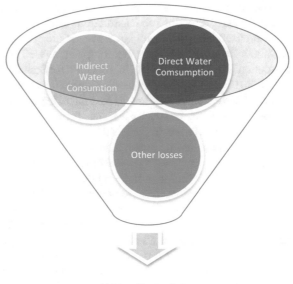

Water Footprints

carry out a local assessment of all sustainability indicators viz. economic, social and environmental impacts; and draw information and consequences regarding sustainable and scrupulous water consumption and its allocation. Therefore, overall it can be concluded that the main objective of analysing water footprints is to find out the relationship of different products or human activities with the problems of scarcity of water and pollution. Further, these figures of water footprints of help to draw and explore new alternatives to make water consumption associated with products and other activities, more sustainable from environmental point of view. In fact the estimation of water footprint is more likely to be dependent on the scope of interest. Water footprints can be calculated for a process, product, consumer, consumer groups, sector, specific location like nation, province, river basin etc.

The complete process of water footprint estimation comprises of four different phases viz. (i) Formation of goals and scope, (ii) Water footprint accounting, (iii) Water footprint sustainability assessment, (iv) Water footprint response formulation. The beginning of water footprint assessment is done by deciding goals and scope means for what specific reasons the assessment is going to be carried out. In water footprint accounting phase, all data are collected depending on the goals and scope finalized in the first phase. After this estimation of water footprints in the view of social, environmental and economic aspects; is carried out in next phase which is water footprint sustainability assessment phase. In the last phase i.e. water footprint response formulation phase, final conclusions, policies; strategies etc. are developed based on the results of previous phase. These all four phases represent the complete process of assessment of water footprints however, it is not essentially required to include all the phases in a study depending on the objective and purpose of the study (Water footprint assessment Manual 2011).

The global water footprints standards have been introduced which comprise a comprehensive directions and instructions to estimate water footprints. These instructions are useful to provide guidance to estimate green, blue and grey water footprints, sustainability assessment of water footprints as well as to use results for policy implementation. ISO (International Organization for Standardization) specify standard ISO 14046:2014 which provides detailed guidelines pertaining to water footprints estimation. These standards allow to find out effective water footprints values and data related to products, processes etc. and further interpret results in the form various impact indicators (www.iso.org). Also, various web-based tools are also available for water footprint assessment which are very helpful for researchers and policy makers to make strategies for improvement of sustainability of water resources and their use.

In hydropower facilities large amount of water is collected in dams to provide smooth and continuous flow for uninterrupted power generation. However, water consumption from dams is not only associated with power generation rather there are several other factors which are also important to consider in water footprint assessment. Figure 3 shows a simple schematic classification of hydropower facilities. Hydropower facilities can broadly classified as large power units and small power units. There is no fixed limit of power output to classify small hydropower and large hydropower units. Every country has decided a limit of output power like 50, 25, 10 MW etc. to differentiate small hydropower units from large units. Large

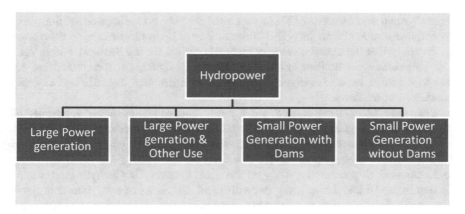

Fig. 3 Classification of Hydropower facilities

power plants can be established for sole power generation only and for multi-purpose (power generation with other uses like irrigation etc.). Small hydropower units are generally deployed for power generation mainly and they can be categorised as run-of-river units, canal bases units and dam toe units. Theoretically, small hydropower units without dams may be considered to have no water consumption. In all dam based hydropower units, the effects of evaporation from reservoir and evapotranspiration process are important in water footprint assessment. Also, it has been seen in many water footprint studies that the non-consumptive share of water withdrawals or the return flow is not considered as a part of the water footprints.

3 Studies Related to Water Footprints of Hydropower Projects

There are many studies pertaining to hydropower water footprints assessment reported till now. Gleick (1992) carried out a study on environmental consequences associated with hydropower projects and concluded that in the cases where gross static head of plants exceeds the dam height; the smaller hydro dams consumes more land and water and its storage per unit power generation than those dams having capacity more than 25 MW.

Herath et al. (2011) presented a study related to comparison of water footprints of hydropower production in New Zealand with three different methods in a case study. These three different methods were based on gross water consumption, net consumptive use and net water balance respectively. The method based on net consumptive use considered the impacts on land use due to establishment of dam. It was found that the water footprints estimates based on first and second method were higher than those based on third method.

Mekonnen and Hoekstra (2012) presented a study on estimation of the water footprints associated with thirty five different sites of hydropower dams. It was found in the study that the average value of water footprints for the dams was 68 m³/GJ. The variations were noticed in the water footprints figures for the studied project due to the large variations in area flooded per unit installed capacity and climatic conditions of the dams.

Zhao and Liu (2015) carried out a study and estimates water footprints of hydropower generation and water footprints of reservoir independently using a new parameter i.e., allocation coefficient which was based on the ratio of power generation benefit to the total ecosystem service benefits.

Lampert et al. presented a study on the water consumption in hydropower generation in the United States using three different categories of hydropower projects viz. run-of-river projects, projects producing power using multipurpose dams and projects producing power using dedicated reservoirs. The results showed negligible water consumption from run-of-river projects; and water consumption associated with dedicated and multipurpose reservoirs are 10.2 and 22.7 gallons/kWh of electricity generation respectively.

Bakken et al. (2016) presented a study on assessment of life cycle water footprints of two hydropower projects in Norway. The water footprints calculated for the projects comprises water consumption for construction and operational phases. The study also found large differences in the values of water consumptive rates associated with hydropower production as compared to other technologies of power production. The reason explained for this was absence of clear guidelines on methodology used for estimation of water consumption.

Scherer and Pfister (2016) have carried out a study to analyse water footprints of 1500 hydropower projects and found that variation in flow regimes may have adverse effects on ecosystem. The methodology used in the study tried to fill the gap of neglecting evapotranspiration before the dam construction. The study concluded that most of the previous studies have shown overestimated impacts of hydropower on scarcity of water.

Strachan et al. (2016) carried out a study to analyse variation in evaporation due to establishment of reservoir for hydroelectric power generation in Canada. The study was based on eddy covariance technique to measure flood landscape water flux before and after the creation of dam over four different ecosystems for a period of five years. The study found the increase of evaporation after the establishment of the reservoir.

Grubert (2016) presented a study to evaluate national and regional data of gross evaporation and net evaporation for hydroelectricity in United States. The study found that net evaporation from reservoir is better than gross evaporation to assess the impacts of hydropower facilities.

4 Discussions

A less number of studies were found exclusively related to the water footprints of hydropower projects. Moreover, a large variation was found in the estimated values of the previous studies. It was found in many studies that the overall gross water consumption was expressed as a ratio of total evaporation loss from a reservoir to annual electricity generation. This looks a quite simple formulation and has limited scope to present the real picture of water footprints of hydropower. Apart from this, some studies have used net water consumption to estimate water footprints which is defined as the ratio of difference between evaporation from the reservoir and evaporation before inundation to the annual power production. This method incorporates the evaporation losses associated with a specific location before the construction of dam. Another approach to calculate water footprints is water balance which is expressed as the ratio of the difference between evaporation from the reservoir and direct rainfall to the reservoir to the annual power production. The lack of a consistent and clear methodology to calculate water footprints of hydropower systems is one of the limitations. The evaporation losses associated with artificial lakes developed by dam reservoirs are often approximated same as those of a natural lake. This can be helpful for estimating some data of water consumption of dams but fails to provide more reliable results to reveal the correct water consumption from hydropower units. So there is a requirement to develop an effective methodology for estimating evaporation losses from an artificial dam reservoir. Also the methodology must be able to incorporate the indirect effects on water use and land nearby due to the construction and creation of the reservoir.

Another important limitation in calculating the water footprints is the lack of deciding system boundaries in space and time more clearly. Table 2 shows some values of water footprints estimated in previous studies. Large variations in the values of water footprints may be noticed and it is quite complex to compare them. The results of any life cycle analysis are greatly depending on the system boundaries decided in a study. In various studies, the spatial boundaries are confined to the catchment of the reservoir and power plant but the establishment of a hydropower facility always affect the downstream side of the reservoir and further beyond the catchment area. The increase of spatial boundaries in a water footprint study makes the analysis complex but it enhances the quality of the results which gives more clear and useful results regarding the real effects of hydropower schemes. Another important thing is the variation of power generation with the variation of the inflow to the reservoirs affects the water consumption figures. Moreover in multipurpose dams the main objective of the reservoir may be different from power generation. The reservoir may useful to serve other purposes like water supply, irrigation etc. even when the power generation is low. These conditions become some important factors which may vary the water footprint estimates with time and seasons. In addition to this, power generation of a hydropower plant is also dependent in other socio-economic and technical factors. In now days the establishment of reservoirs dams is also being associated with the climate change issues. So, it becomes important to

Table 2 Water footprints estimation in previous studies

S. No.	Author	Country	Water footprints
1	Herath et al. (2011)	New Zealand	1.55–6.05 m^3/GJ
2	Mekonnen and Hoekstra (2012)	Global	68 m^3/GJ (avg.)
3	Zhao and Liu (2015)	China	3.6 m^3/GJ
4	Lampert et al.	USA	10.2–22.7 gallons/kWh
5	Bakken et al. (2016)	Norway	0.15 (l/kWh)
6	Scherer and Pfister (2016)	Global	55 m^3 H_2Oe/GJ
7	Grubert (2016)	USA	1.7 m^3/GJ

consider the past years data of evaporation and power generation while estimating the water footprints of the upcoming years. The water footprints studies are mainly focuses on estimation of water consumption in operational phase of the reservoir. The inclusion of water footprints during other phases like construction, demolition, maintenance phases of the reservoirs further improves the results. It can be seen in other previous life cycle studies related to hydropower which focuses on estimating life cycle GHG emissions; acidification potential, energy etc. have considered the construction phase and even demolition phase.

In almost all situations, reservoirs themselves may be helpful in controlling various impediments. A large amount of water stored in hydropower dams is capable to control water scarcity problem nearby. As already discussed that multi-purpose dams are established to serve other purposes like irrigation, water supply, environmental flow, flood protection, fisheries etc. apart from power generation. Another important thing is to connect the water footprints assessment to the global level rather than local. In almost all cases, water footprints analysis is basically concerned with the water resources of a specific region and territory. Even if it is somehow complex to relate the water footprints assessment to global hydrological cycle but in many cases it is essentially required to popularise as an important environmental impact in all life cycle assessment (LCA) studies. This is because the main objective of water footprints assessment to manage and control the water scarcity problem in future as well as to make the current technology more sustainable in water perspective. Therefore, new methodologies must be effective to incorporate all above discussed factors including the advantages of large amount of water availability in dams and estimating reliable and unbiased results to study the hydropower facilities.

5 Conclusion

The chapter presents an overview of water footprints assessment of hydropower projects. The electricity generation capacity through hydropower facilities is increasing at rapid rate. But at the same time environmental concerns associated with hydropower projects are coming up too. Hydropower facilities use large water storage reservoirs to generate power and other social uses. Studies are being carried out to estimate water footprints associated with hydropower. From the previous studies, it was found that there is still lack of a common and effective methodology to estimate water footprints of hydropower dams. It is important to incorporate other uses of hydropower dams other than hydropower while estimating water footprints. The inclusion of more factors while calculating hydropower footprints would help to explore a clear and unbiased picture of sustainability of hydropower projects in future. In addition to this, more realistic calculations would be helpful to compare large and small hydropower projects from economic and environmental point of view. An effective and clear methodology may further help to develop some generalised mathematical correlations for water footprints assessment.

References

Bakken, T. H., Killingtveit, A., Engeland, K., Alfredsen, K., & Harby, A. (2013). Water consumption for hydropower plants-review of published estimates and an assessment of the concept. *Hydrological and Eath System Sciences, 17,* 3983–4000.

Bakken, T. H., Modahl, I. S., Engeland, K., Raadal, H. L., & Arnoy, S. (2016). The life cycle water footprint of two hydropower projects in Norway. *Journal of Cleaner Production, 113,* 241–250.

Gleick, P. H. (1992). Environmental consequences of hydroelectric development: The role of facility size and type. *Energy, 17*(8), 735–747.

Grubert, E. A. (2016). Water consumption from hydroelectricity in the United States. *Advances in Water Resources, 96,* 88–94.

Herath, I., Deurer, M., Horne, D., Singh, R., & Clothier, B. (2011). The water footprint of hydro-electricity: A methodological comparison from a case study in New Zealand. *Journal of Cleaner Production, 19,* 1582–1589.

Hoekstra, A. Y., & Chapagain, A. K. (2007). Water footprints of nations: Water use by people as a function of their consumption pattern. *Water Resource Management, 21,* 35–48.

Hoekstra, A. Y. (2003). Virtual water trade: Proceedings of the international expert meeting on virtual water trade, December 12–13, 2002. Value of Water Research Report Series No. 12. Delft, Netherlands: UNESCO-IHE.

Lampert, D. J., Lee, U., Cai, H., & Elgowainy, A. *Analysis of water consumption associated with hydroelectric power generation in the United States* (greet.es.anl.gov). Assessed on April 13, 2018.

Mekonnen, M. M., & Hoekstra, A. Y. (2012). The blue water footprint of electricity from hydropower. *Hydrology and Earth System Sciences, 16,* 179–187.

Renewable. (2017). Global status report.

Scherer, L., & Pfister, S. (2016). Global water footprint assessment of hydropower. *Renewable Energy, 99,* 711–720.

Strachan, I. B., Tremblay, A., Pelletier, L., Tardif, S., Turpin, C., & Nugent, K. A. (2016). Does the creation of a boreal hydroelectric reservoir result in a net change in evaporation? *Journal of Hydrology, 540,* 886–899.

Varun, Ravi Prakash, & Bhat, I. K. (2009). Energy, economics and environmental impacts of renewable energy systems. *Renewable and Sustainable Energy Reviews, 13,* 2716–2721.

Varun, Ravi Prakash, & Bhat, I. K. (2010). Life cycle energy and GHG analysis of hydroelectric power development in India. *International Journal of Green Energy, 7,* 1–19.

Varun, Ravi Prakash, & Bhat, I. K. (2012). Life cycle greenhouse gas emissions estimation for small hydropower schemes in India. *Energy, 44,* 498–508.

Water footprint assessment Manual (2011). *Water footprint network.* www.waterfootprint.org. Assessed on April 13, 2018.

www.iso.org. Assessed on May 08, 2018.

www.waterfootprint.org. Assessed on April 13, 2018.

Zhao, D., & Liu, J. (2015). A new approach to assessing the water footprint of hydroelectric power based on allocation of water footprints among reservoir ecosystem services. *Physics and Chemistry of the Earth, 79,* 40–46.

An Empirical Investigation into Water Footprint of Concrete Industry in Iran

S. Mahdi Hosseinian and Reza Nezamoleslami

Abstract Reduction of water consumption of concrete production is of particular importance within the construction industry to take steps toward sustainable construction materials. However, a lack of available benchmark metrics has made it difficult for governments to identify areas to target for water consumption reduction and even to provide a basis to analyse water consumption impacts of concrete production on their national environment. This chapter looks at water footprint of concrete industry based on a life cycle assessment approach. A comprehensive water footprint model of concrete production is provided. Elaboration is paid to the raw materials (cement and aggregates), energy, transportation and human's food, as important factors affecting water footprint of the concrete industry. A large cement plant, a concrete plant and an aggregate producer in Iran are analysed and effects of different parameters on the water footprint model are evaluated based on a sensitivity analysis method. The chapter shows that the water consumption intensities of cement, aggregate and concrete productions account for 2.126, 0.583 and 0.967 m^3/ton, respectively demonstrating that the concrete industry should be treated as a high water consumer industry. The chapter demonstrates that shifting to a high contribution of renewable energy is one effective solution for the water consumption problem of the concrete production. In addition, the chapter illustrates that the personnel's food contributes to 6850, 565,000, and 22,610 m^3 water footprint, in the investigated concrete, cement and aggregate plants in 2017, respectively; showing the effect of human management on the water footprint reduction. This chapter will be of interest to those seek sustainability in concrete production.

Keywords Water footprint · Concrete production · Water consumption

S. Mahdi Hosseinian (✉)
Department of Civil Engineering, School of Engineering, Bu-Ali Sina University, Hamadan, Iran
e-mail: s.hosseinian@basu.ac.ir

R. Nezamoleslami
Master of Civil Engineering, Bu-Ali Sina University, Hamadan, Iran
e-mail: rezaci.nezami@gmail.com

© Springer Nature Singapore Pte Ltd. 2019
S. S. Muthu (ed.), *Environmental Water Footprints*, Environmental Footprints
and Eco-design of Products and Processes,
https://doi.org/10.1007/978-981-13-2739-1_3

Nomenclature

EWR   Environmental water requirements
DWC   Direct water consumption
ISO   International Organization for Standardization
LCA   Life cycle assessment
MAR   Mean annual runoff (total amount of available fresh water)
VWC   Virtual water consumption
WCI   Water consumption intensity
WF   Water footprint
WSI   Water stress index

1 Introduction

Environmental issues were historically seen as a secondary priority or something compulsory and commonly considered as being against to revenue within the construction industry (Carmichael and Balatbat 2009). Fortunately, this view is now changing and sustainability now contributes to the well-being of construction companies, material manufacturers, and construction personnel. Such sustainability perception has particularly been stimulated recently in the built environment' arena (Kibert and Fard 2012). This has encouraged a number of emerging issues in sustainability as it impacts the practice of construction along with the production of construction materials (Siew 2015).

Concrete as one of the most common construction materials plays a central role in the construction industry. Due to the increased demand for infrastructure, concrete usage increasingly continues to rise (Sjunnesson 2005; Sakai and Noguchi 2013; Georgiopoulou and Lyberatos 2017). However, the production of concrete is characterized by water consumption and emission production (Sjunnesson 2005). Addressing these environmental problems is considered as one important step toward sustainable construction (Zhang et al. 2017). Emission production problems of concrete have long been recognized and research in this area is well developed (Gartner 2004; Josa et al. 2004, 2007; Georgiopoulou and Lyberatos 2017). However, there has been very little work directed towards water consumption problems of concrete production. Due to the global water scarcity there is a growing concern for concrete industry to seek strategies to reduce its water use (Horvath and Matthews 2004; Huntzinger and Eatmon 2009).

Water shortage is now seen as a major challenge of the concrete and cement industry in dry countries (Doe 2006; Mielke et al. 2010). For instance, Iran is located in arid and semiarid regions (Ababaei and Etedali 2017) and it suffers from water shortage problems which have threatened Iran cement and concrete industry (Chehreghni 2004). Iran relies on cement production as it equals to 0.8% of its gross domestic product. After China, India and the USA, Iran is ranked four in terms of cement production with around 76 million ton cement production annually (Bod 2014). This has raised an inevitable trade-off between environmental protection and development in Iran.

Unfortunately, limited published data have made it difficult for the Iranian government and relevant organizations to able to seek strategies and promoting technologies required reducing water consumption. Hosseinian and Nezamoleslami (2018) and Mack-Vergara and John (2017) claim that water consumption data of concrete production and its raw materials have not well documented. Therefore, studies concerning water consumption of concrete production should be encouraged. It seems that concrete organizations have been reluctant to make data for fear of identifying inefficiencies within their concrete products (Acquaye et al. 2017). Finnveden et al. (2009) and Pfister et al. (2011) argue that industrial water consumption data is limited.

In such light, this chapter looks at water footprint (WF) of concrete industry through adopting a life cycle assessment (LCA) framework. A comprehensive WF model of concrete production is proposed. Elaboration is paid to the raw materials (cement and aggregates), energy, transportation and human's food, as important factors affecting WF within the concrete industry. A large cement plant, a concrete plant and an aggregate producer in Iran are analysed and a sensitivity analysis is conducted. To reduce the water consumption of the concrete production recommendations are provided.

Aims

On having worked through this chapter, the reader should be able to appreciate:

- The WF methods that might be used for concrete WF studies.
- The main contributing factors to the WF of concrete.
- The WF amount of concrete industry.
- The ways that can be used to reduce the WF of the concrete industry.

Structure

The chapter is structured in nine sections:

- Background.
- Methodology
- Proposed concrete WF model
- Empirical studies
- Discussion
- Sustainability assessment
- Sensitivity analysis
- Model validation
- Conclusion.

This chapter provides a benchmark to assess the WF of the concrete industry; thereby the chapter contributes to the environmental management of this industry.

2 Background

Freshwater, as the most important natural resource on earth, merely constitutes to 2.53% of the whole water on Earth. However, the majority of the fresh water is frozen in glaciers and polar ice caps and is usable for humanity (Shiklomanov and Rodda 2003). The fresh water availability varies around the planet, and unfortunately dry and populated areas are most exposed to water shortage problems. Typically water shortage problems are expressed by terms such as water stress and water scarcity. The term water stress is used here in the sense of Schulte (2014) to refer to the ratio between used water and available water. The term water scarcity is used in the sense of Schulte (2014) to refer to a lack of sufficient freshwater to encounter the water usage requirements of an area. Water stress and scarcity problems, with their associated impacts, are becoming now issues of increasing concern across the globe. For example, the water consumption for energy generation is predicted to rise 60%, in 2050 (Bruinsma 2009; UN 2013). Such water demand rise necessitates the need for effective management of the freshwater resources in future to effectively manage growing water stress and scarcity. In this regard, traditional methods for water management are now implemented with a more advance tools and concepts. Investigating WF of products and their associated environmental impacts are marked as part of such advancements.

The WF concept was firstly mentioned by Allan (1998) when the term virtual water was introduced for identifying that the water consumption of producing citrus fruit is considerably larger than the water content of the fruit itself. Virtual water represents the water polluted or used somewhere else to make the product (Allan 1998; Chapagain and Orr 2009; Verma et al. 2009; Gao et al. 2011). In 2002, Hoekstra developed the WF concept for different kinds of water polluted or consumed (Hoekstra 2002). Such concept is now of particular importance in the sustainability schedule of many countries (Mekonnen and Hoekstra 2014; Gu et al. 2015; Zhuo et al. 2016) and it is now applied for the analysis of water consumption of different industries at the scales of global, national, and regional (ISO 2014). Assessing the WF of a product gives information required by manufacturers to evaluate their need on water rare resources (Chapagain and Orr 2009; Ridoutt and Pfister 2010).

Although the WF term is well understood in many industries, it is not yet a clear established topic in the concrete industry (Hoekstra et al. 2011; Hoekstra and Mekonnen 2012). A number of studies have been conducted concerning WF of products, such as steel (Gu et al. 2015), paper (Van Oel and Hoekstra 2012), food and beverage (Ercin et al. 2011, 2012), and fibre (Chico et al. 2013); however, few studies have looked at WF of concrete and its raw materials (aggregates and cement).

A number of researchers have attempted to measure the direct fresh water consumption of concrete industry (for example, Chehreghni 2004; Chen et al. 2010; Lafarge 2012; Valderrama et al. 2012; Cemex 2015; Holcim 2015). The term fresh water is adopted here in the sense of Huntzinger and Eatmon (2009) to refer to groundwater, fresh tap or surface water used for a product manufacturing. The term water consumption is used here in the sense of Williams and Simmons (2013) to refer to the amount of abstracted water which is not returned to the groundwater or surface in the same drainage basin where it was withdrawn. Used water might be incorporated into products, transpired, evaporated, or otherwise removed. There is a distinction between used and consumed water. Used water may or may not return to its source again while consumed water cannot be returned. Valderrama et al. (2012) argue the effect of the cement line machinery on the direct water use. The study of Chen et al. (2010) looks at the direct water consumption intensity of cement production in France and estimates its value as 0.200 m^3/ton. The studies of Lafarge (2012), Holcim (2015), and Cemex (2015) give figures of 0.185–0.808 m^3/ton for the direct water consumption intensity of cement production. Water intensity refers to the proportion of the amount of water consumed or withdrawn to the unit of product which is produced (Williams and Simmons 2013). Water intensity includes reclaimed and recycled water which mostly be influenced by the production's process and equipment usage (Huntzinger and Eatmon 2009). Cemex (2015), Holcim (2015) and Lafarge (2012) estimate the direct water consumption intensity of concrete aggregates between 0.116 and 0.413 m^3/ton, and that of concrete production between 0.2 and 0.25 m^3/ton.

Although measuring direct water consumption provides a useful view regarding water consumption of concrete production, it fails to give information about virtual water consumption of concrete production. To consider both direct and virtual water consumption of concrete production the WF concept needs to be adopted. As mentioned before, few researchers have looked at the WF of concrete production. Available data is generally concerned with carbon footprint and energy consumption of concrete industry (Hasanbeigi et al. 2012). Mack-Vergara and John (2017) have published a concrete WF assessment paper, although their focus was on a hypothetical concrete production scenario rather than a real world data gathering. Netz and Sundin (2015) investigated the WF of concrete production in Sweden, and in the same vein Mellor (2017) looked at the WF of concrete production in New Zealand. No research has investigated WF of concrete in a dry country and there are few WF studies using Iran data. Hoekstra and Mekonnen (2012) argue the importance of scale national level data in WF investigations. This chapter is an attempt to address such knowledge gap.

Various tools and indicators may be adopted for assessing and benchmarking of a product's (concrete here) environmental problems, such as life cycle assessment (LCA), environmental impact and risk assessment, strategic environmental assessment, feasibility analysis, and ecological footprint (Finnveden and Moberg 2005; Ness et al. 2007). By contrast, this chapter only focuses on the LCA of concrete production. The WF concept is applied to different water types, namely, blue, green and grey (Gu et al. 2015). Blue WF is the focus of this chapter as it refers to ground and surface water which are withdrawn from the environment for human requirements. Green water is defined as precipitation and grey water refers to the amount of fresh water used to dilute concentrations in polluted water, caused by a certain process step, back to its natural water quality (Hoekstra et al. 2011).

3 Methodology

Different methodologies might be utilized to account for the WF of products for example ISO 4046 (ISO 2014) and the Water Footprint Network, WFN (Hoekstra et al. 2011). In this chapter the WFN methodology, due to its popularity is adopted. For the development of a feasible WF methodology a LCA approach needs to be adopted (Jeswani and Azapagic 2011). Therefore, the concept of LCA is first discussed.

3.1 Life Cycle Assessment

A LCA approach is utilized to evaluate the environmental impact, carbon emission, acidification, eutrophication or water use, during the whole life cycle of a product from raw material acquisition to waste management, known as cradle-to-grave, (Finnveden et al. 2009). LCA approaches generally comprise four steps (Hunt et al. 1992; Owens 1997; ISO 14040 and 14044 2006a, b; Huntzinger and Eatmon 2009). The first step, definition of goal and scope, specifies the subject of the study, model boundaries and required data. The second step, life cycle inventory, LCI, evaluates inputs (for example raw materials) and outputs (for example water use) throughout the lifetime of the product. This step is followed by the life cycle impact assessment step which comprises identifying, quantifying and evaluating the results of the LCI. In the final step, interpretation of the results and recommendations are provided.

3.2 Water Footprint Measurement Based on Water Footprint Network

To measure WF of a product, the Water Footprint Network (WFN), a non-profit organization, seeking efficient water use and sustainability, have developed a manual called Water Footprint Assessment Manual (Hoekstra et al. 2011). The manual contains a number of definitions and methods for WF assessment. It looks at WF of products, processes, consumers, nations and businesses.

Four different stages are considered in the lifetime of a product based on the scope of the WF assessment (Hoekstra et al. 2011; Herath et al. 2011b). Firstly, goal and scope are defined similar to the LCA approach. The goal can be to examine the WF of a product or a process. A scope might describe what to include and exclude in the assessment. The result of the assessment might vary depending on limitations considered in this step. The options to include blue, green and/or grey water are also considered in the first step. In the second step, which accounts for WF calculation, required data are collected and processed. In this step, the WF calculation might be divided to direct or indirect consumed water in the different processes of the supply chain of the product. Direct water usage includes the water consumed in a specific

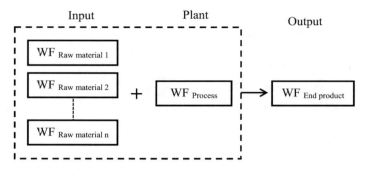

Fig. 1 WF calculation based on water footprint network approach

process while indirect water usage refers water used in preceding processes. The calculation is typically made separately for blue, green and grey water. The WF of a product (output) is calculated by simply summing the WFs of the raw materials (input products) and the process, as shown in Fig. 1. That is the WF of a product is obtained by considering the raw resources (where the supply chain begins) and then calculating, step-by-step, the WFs of the middle products, until the WF of the end product is obtained. Hoekstra et al. (2011) argue that such calculations are considered as main barriers for studies concerning WF of a product as they require a huge data gathering which is mostly not accessible.

In the third step of the WFN approach, sustainability assessment of the WF is made. This is primarily considered by comparing the actual water consumed by a product with the fresh water available in the region where water is withdrawn. The final step of such approach provides strategies to reduce WFs.

3.3 Model Boundary

The life cycle of the concrete production comprises supply of raw materials (largely cement and aggregates), concrete manufacturing, concrete usage, demolition, and transportation. Accordingly a LCA needs to be adopted to evaluate the effects of concrete on aquatic environments over the whole concrete life cycle. However, WFs of concrete product consumption (e.g. for placing, curing and maintenance) vary considerably based on the end usage (e.g. buildings, dams, and pavements) and are difficult to obtain. The impact from the manufacture of machineries used in the different processes of concrete production is not significant and is typical not considered. In addition, the WFs of the concrete admixture can be very diverse based on the sources and are hardly documented. Given a low proportion of additives, this is likely a small portion of the overall WF. Finally, the water used in the demolition of the end product and the use of demolition products are not typically tracked.

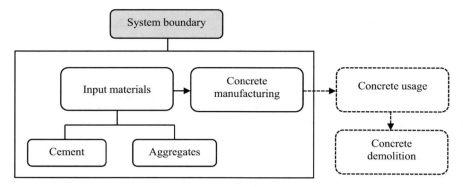

Fig. 2 System boundary for assessing water footprint of concrete industry, the arrows show transports

Figure 2 illustrates the system boundary (what is considered in this chapter is in solid lines). The cement and aggregate supply and the concrete manufacturing are treated as the most important body which are seen as the main part that producers need to consider when deciding on reducing water consumption of concrete production. Comparing Figs. 1 and 2, the water consumed in the concrete plant is considered as the WF of the concrete processing and the WF of the cement and aggregates are treated as the WF of the input materials. Thus in the following, the WF assessment models of cement and aggregate production along with concrete manufacturing are proposed.

3.4 Cement Production Phases

Cementitious materials are generally regarded as inorganic materials mixed with water and used to bind other materials together. Hydraulic cements, with the special property of setting and hardening under water, are the most popular types of cementitious materials. The focus of this chapter is Portland cement because it is regarded as the most common hydraulic cement. Typically, Portland cement is produced form clay and limestone through heating it in a rotary kiln (Neville 1995). Production of Portland cement includes mixing of raw materials through implementing a dry or wet process, burning, grinding, storage as well as packaging (Schneider et al. 2011). This chapter considers the dry process as it is the popular form for producing cement.

The life cycle of cement comprises raw materials (mainly lime clay and stone) acquisition, cement manufacturing, transportation, cement consumption, and cement products' disposal and recycling (Schneider et al. 2011). Figure 3 shows the cement production phases. In each phase a huge amount of water and energy is required (Huntzinger and Eatmon 2009). Thus complete LCA and inventory analyses might be rather complex. In practice, WFs of raw materials (clay, limestone) of cement are

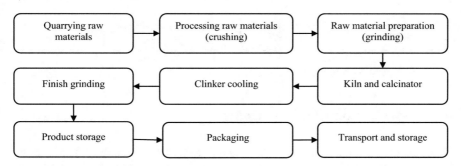

Fig. 3 Cement production phases

often difficult to obtain as the extraction and transportation of raw materials might be very diverse and depend on their resources. Also the water consumed for the construction and demolition of the cement plant is not well documented. Owing to the multi-decade life of most cement plants, this is expected to be a small fraction of the whole WF.

3.5 *Aggregate Production Phases*

Aggregate occupies 60–75% of the volume of concrete, so its WF is of great importance. Aggregates might be classified according to their size to fine (sand) and course (gravel). Fine aggregate is smaller than 5 mm and accounts for 35–60% by mass or volume of total aggregate. Course aggregate is bigger than 5 mm and typically is between 9.5 and 37.5 mm. Aggregates might also be classified based their shape to rounded and crushed. Rounded aggregates are dug from rivers, lakes and seabed while crushed aggregates consist of crushed quarry rocks. Due to the environment protections, in recent years, it has become common to use crushed rather than rounded aggregates (Netz and Sundin 2015). This is largely because rounded aggregate is necessary in rivers as it purifies the groundwater to achieve good quality. Crushed aggregate is the focus of this chapter.

Aggregate production phases involve raw material extraction, washing, material classification, transportation and storage. Water is consumed in each phase and the amount of water consumption depends on different parameters such as extraction process, sources, equipment used and types of energy consumed (Korre and Durucan 2009; World Business Council for Sustainable Development 2014). Such parameters make calculating WF for aggregate production highly difficult. Figure 4 presents aggregate production phases.

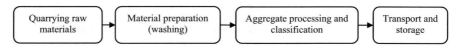

Fig. 4 Aggregate production phases

4 Proposed Concrete WF Model

Based on the WF concept (direct and virtual water), system boundary, demonstrated in Fig. 2, and the WFN approach, Fig. 5 shows the WF model for concrete production and Fig. 6 presents the WF model for its input materials (aggregate and cement). As discussed earlier, the cement and aggregate supply and concrete production process are regarded as the key discussion of the chapter, which is expected to be the central part that manufacturers need to pay attention when they deal with water risk in industrial WF evaluation.

It follows directly form Figs. 5 and 6 that the WF amount can be calculated by,

$$WF = VWC + DWC \tag{1}$$

where VWC and DWC, respectively, represent the virtual and direct water consumptions.

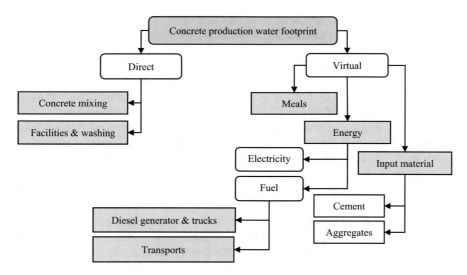

Fig. 5 WF model for concrete plants

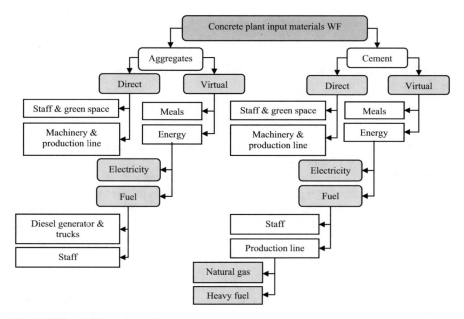

Fig. 6 WF model for concrete input materials

4.1 Direct Water Consumption (DWC)

The proportion of direct water use differs among different plants given the plant capacity, water used for facilities, technology and machinery utilized, and production process (Huntzinger and Eatmon 2009). Direct water consumption also varies based on the concrete mix deign (slump value, water content) and type of concrete (normal, special). In the proposed model the amounts of direct water used for facilities, personnel, concrete mix, and trucks and equipment washing are considered as the direct water consumption of concrete production.

4.2 Virtual Water Consumption (VWC)

The proportion of virtual water consumption can be obtained by simply summing the WFs of different parameters which consume virtual water,

$$\mathrm{VWC} = \sum_{i=1}^{n} \mathrm{VWC}_{P_i} \tag{2}$$

where VWC_{pi} can be calculated by,

$$\text{VWC}_{P_i} = A_{P_i} \times \text{WCI}_{P_i} \tag{3}$$

In Eq. (3), A_{P_i} is the amount of the each parameter used in the concrete production and WCI_{P_i} is its corresponding water consumption intensity. The WCI_{P_i} value can be obtained by,

$$\text{WCI}_{P_i} = \text{WF}_{P_i}/\text{Prod}_{P_i} \tag{4}$$

where WF_{P_i} and Prod_{P_i} represent the WF and amounts of production of parameter i in a specified period of time, respectively.

Based on Fig. 5, the parameters which consume virtual water comprise input materials (cement and aggregates), energy sources used in the concrete plant and the personnel food. Among these parameters the WFs of the energy sources have been discussed in the relevant literature; however, there is a lack of research on the WF of cement and aggregates. So in this chapter the WFs of cement and aggregate plants are investigated through an empirical study.

Energy water footprint

There is a close link between water and energy to make sure a sustainable supply of each of them. Recently, the link between water and energy has motivated many researchers (Scown et al. 2011). The selection of energy is generally dependent on its cost and availability. The use of different energy sources varies in each country. In Iran, oil, natural gas and diesel are main preferred energy types for heat and electricity production. Therefore, in the proposed models these three fuel resources are discussed.

The WF of natural gas is typically defined based on water volume per unit of volume (m^3/m^3), though WF might be expressed in terms of water volume per unit of embedded electricity or heat energy (m^3/TJ). Using the data presented in Enerdata (2015) and Mekonnen et al. (2015), the research of Hosseinian and Nezamoleslami (2018) calculates the WCI of natural gas as 9.251 L water per cubic meter of natural gas.

Water is used in the extraction and refining of crude oil. A barrel of crude oil is 159 L and from refining it around 3.8 L of heavy fuel oil can be obtained. Williams and Simmons (2013) show that for the extraction of one liter of heavy fuel oil (equivalent to 41.84 L of crude oil), 22.5 L water and for refining it between 8.25 and 40 L water are required. Therefore, WCI for the extraction and refining one liter of heavy fuel oil can be obtained from 30.75 to 62.5 L.

It can be shown that one liter refined diesel fuel is equivalent to 3.82 L of crude oil (Williams and Simmons 2013). Accordingly, WCI for the extraction and refining one liter of diesel fuel can be estimated between 2.81 and 5.62 L (Hosseinian and Nezamoleslami 2018).

For gasoline-based (diesel-based) vehicles, King and Webber (2008) estimate water use in average between 0.16 L (0.18 L) and 0.33 L (0.26 L) water/km, respectively.

Table 1 Water consumption intensity data

Natural gas (L/m^3)	9.251
Heavy fuel oil (L/L)	30.75–62.50
Diesel fuel (L/L)	2.81–5.62
Gasoline for transportation (L/km)	0.16–0.33
Diesel for transportation (L/km)	0.18–0.26
Electricity (L/MWh)	1800
Meals (L/meal)	4756.88

Using the findings of study of Mekonnen et al. (2015), it can be shown that the water used for electricity generation in Iran is equivalent to 0–1.8 m^3 water per megawatt hour (MWh).

Employees water footprint

In a WF study, Hosseinian and Nezamoleslami (2018) suggest to calculate the WF of employees based on the WF of their consumed food during their working time. Inspiring idea from food ecological footprint (Spiess 2014) they propose a novel method to calculate the WF of employees based on food market price, the cost contribution of food items of the meal, the cost of each item of the food, and WF of each food item of the meal. Based on information provided by Domenech Quesada (2007), Hoekstra (2008), Ercin et al. (2011), and Spiess (2014) they calculate WF per a typical meal as 4756.88 L.

4.3 Summary of WCI

Following the above discussion, Table 1 gives the water consumption intensity for the parameters considered in this chapter.

5 Empirical Studies

In this section, three cases, a large cement plant, an aggregate producer and a concrete manufacturer, located in Iran, are evaluated and the application of the model is provided.

5.1   Cement Plant

The selected cement plant, producing Portland cement, has the annual capacity of 1.7 million ton cement with 300,000 m^3 annual direct water consumption. Around one percent of the direct water consumption of the cement plant is wasted. Hence after excluding one percent water waste, the amount of DWC is estimated to be 297,000 m^3. The discharge of water of the cement plant is small and can be ignored. Table 2 demonstrates the amount of different energy types used in the selected cement plant with their associated WF values. The information of WCI for calculating VWC in Table 2 is obtained from Table 1.

According to Table 2 the total VWC of the energy consumed in the selected cement plant in one year is estimated to be 2.752 × 10^6 m^3.

Based on the collected date, approximately 430 employees work at the cement plant and the majority of them travel about 15 km from the nearest city to the plant. Daily transport services including 18 vehicles with 20 people capacity are used for carrying the employees from their home to the plant and return them. The vehicles are diesel based and they in average travel 160,000 km per year for the employees' transportation. Using Table 1, the VWC of employees' transportation can be calculated (28.8–41.6 m^3 per year).

The total working time of all employees is 950,000 h in one year. One meal is considered to be eaten by employees in each shift work (8 h). Having this information and using Table 1, the total VWC of the employees' food can be obtained (565,000 m^3).

Figure 7 summarizes the WF calculation results for the selected cement plant.

Figure 7 shows that the WF of the cement plant, with 1.7 million ton cement production per year, is expected to be 3.614 million m^3 for one year indicating that the plant poses a huge risk to the water resources. The results estimate WCI and direct water consumption as 2.126 and 0.175 m^3/ton, respectively. The direct water consumption amount is consistent with the research of Lafarge (2012), Cemex (2015) and Holcim (2015) which give figures of 0.185–0.808 m^3/ton for direct water consumption of cement production.

Table 2 Energy usage and VWC of energy of the cement plant in one year

Energy	Energy usage	Average VWC of energy usage
Electricity	1.8 × 10^5 m^3	3.2 × 10^5 m^3
Heavy fuel oil	2.1 × 10^4 m^3	9.8 × 10^5 m^3
Natural gas	1.57 × 10^8 m^3	1.452 × 10^6 m^3
Total amount	–	2.752 × 10^6 m^3

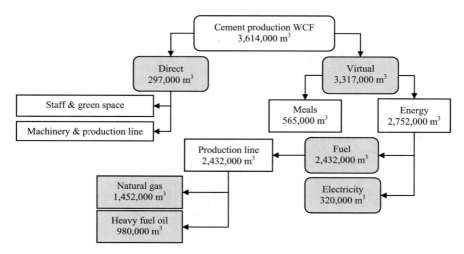

Fig. 7 WF results for cement plant

5.2 Aggregates Plant

The selected aggregate plant, producing crashed aggregates, supplies aggregates required by the concrete plant. The aggregate plant annual capacity is 60,000 tones with 12,120 m³ annual direct water consumption. The direct water waste is assumed to be one percent. Accordingly, the amount of DWC after excluding one percent wasted water equals 12,000 m³. The energy source in this plant is electricity and the fuel type of the machinery is gasoline. Electricity is supplied from the national grid with 102 MWh usage per year and the annual amount of the gasoline usage is 72,000 L. Using Table 2, the VWC of electricity in the aggregate plant is equivalent to 183.6 m³ per year and that of diesel fuel is between 202.32 and 404.64 m³ per year which is averaged at 303 m³ per year. As the plant employees are local the energy usage for personnel transportation is not considered.

The total working time of all employees is 38,000 h in one year according to the information collected from the plant. By using Table 1, the VWC of the employees' food can be calculated (22,610 m³).

Figure 8 summarizes the WF results for the selected aggregate plant.

Figure 8 shows that the WF of the aggregate plant, with 60,000 tonnes production per year, is equal to 35,097 m³. The results show that the amount of VWC (23,097 m³) is 1.9 times larger than that of DWC (12,000 m³). While the direct water intensity of the aggregate plant is calculated 0.2 m³ per each ton aggregate production this chapter demonstrates that by using the proposed WF model the WCI of the aggregates in the selected plant is equivalent to 0.583 m³/ton. This result shows that WCI of the aggregate production is about 3 times bigger than its direct water intensity. The direct

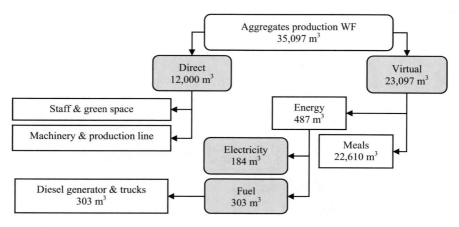

Fig. 8 WF results for aggregate plant

Table 3 Water consumption intensity for concrete plant input materials

Cement (m³/ton)	2.126
Aggregates (m³/ton)	0.585

water intensity result (0.2 m³/ton) is consistent with the data produced by Lafarge, Cemex and Holcim giving figures between 0.116 and 0.413 m³/ton (Lafarge 2012; Cemex 2015; Holcim 2015).

5.3   Summary of WCI for Concrete Plant Input Materials

Based on the above case investigations, Table 3 summarizes the WCI for the cement and aggregates production. The information of Tables 2 and 3 are used to calculate the WF of concrete production presented in the next section.

5.4   Concrete Plant

The annual capacity of the selected plant is 50,000 m³ concrete with 11,000 m³ annual direct water consumption (for facilities, washing and concrete mix). Based on the data gathered the average water intensity for concrete mix in this plant is 190 L/m³. One percent of the direct water is assumed to be wasted. Around 19,500 ton cement is used annually in this plant. For the transport of such cement 488 trucks with 40 ton capacity are required. The distance between the cement and concrete plants

Table 4 Data of case concrete planet in 2017

Direct water consumption	11,000,000 L
Cement consumption	19,500 ton
Gravel consumption	41,500 ton
Sand consumption	55,000 ton
Diesel consumption for diesel generator	46,080 L
Diesel consumption for trucks and loader	43,200 L
Personnel working time	11,250 h
Input materials delivery distance	170,000 km

is 100 km. Thus the cement trucks should travel 200 km for each delivery (return is included). It hence follows that all trucks travels can be calculated as 97,600 km per year.

The annual aggregates (sand and gravel) usage of the selected concrete plant is 96,500 ton (55,000 ton sand and 41,500 ton gravel). For transportation of such aggregates 2413 trucks with 40 ton capacity are required for one year. The distance between the aggregate and concrete plants is 15 km. Accordingly, the aggregate trucks should travel 30 km for each delivery (return is included) and all trucks travels thus become 72,390 km per year. Total travels of cement and aggregates trucks are 170,000 km. As the number of the concrete plant personnel is low and they are local the WF of the personnel transportation is not considerable and is ignored in the calculations.

A 250 KVA diesel generator is used to provide electricity required by the concrete production line. This generator needs 160 L gasoline per day. A wheel loader and a backhoe are used in the concrete plant which they consume 150 L gasoline per day. Five people work 6 days per week at the selected concrete plant (8 h each day).

Table 4 summarizes the data gathered from the concrete plant.

The DWC of the selected concrete plant after excluding one percent water waste (110 m^3) is equal to 10,890 m^3 per year. By using Tables 3 and 4 the VWC related to the input materials (cement and aggregates) of the concrete plant can be obtained $(41,457 \text{ and } 56,453 \text{ m}^3, \text{respectively})$.

The diesel generator consumes 46,080 L gasoline per year. Using Table 2, the VWC of the diesel fuel can be calculated $(194.227 \text{ m}^3 \text{ per year})$. Similarly the VWC of the loader and backhoe fuels can be obtained $(182.082 \text{ m}^3 \text{ per year})$. Accordingly, in the selected concrete plant the VWC of the gasoline fuel is equivalent to 376.315 m^3 per year. Using Table 2 for 170,000 km transportation of the input materials (cement and aggregates), 37.4 m^3 virtual water is consumed in one year. Having the personnel working time per year (11,520 h) and the WCI of one meal (4756.88 L), the VWC of the personnel meal can be obtained (6850 m^3).

Based on the above calculations the WF of concrete production is estimated to be 116,064 m^3 per year and the WCI of concrete production is equivalent to 2.321 m^3/m^3 $(0.967 \text{ m}^3/\text{ton})$. Figure 9 illustrates the corresponding results.

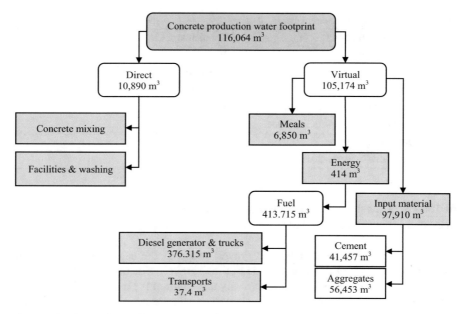

Fig. 9 Concrete production WF

6 Discussion

Direct water consumption of concrete production has been largely discussed in the literature while little attention has been given to the virtual water consumption of the concrete industry. Figure 10 compares DWC and VWC in case studies. This figure shows that VWC contributes to the 92, 66 and 91%, respectively, of the total WF of the selected cement, aggregate and concrete plants. This result shows that a huge water consumption proportion is associated with virtual water which stands as a key point of the water consumption assessment.

Figure 11 demonstrates the WF of different parameters in the selected concrete plant. This figure shows that input materials (aggregates and cement) account for 97,910 m³ WF contributing to the 84.36% (aggregates 48.64% and cement 35.72%) of the total WF of the concrete production. This highlights the role of cement and aggregate on the concrete WF reduction. Although the WCI of cement is approximately 3.5 times larger than that of aggregates a higher contribution of aggregates to the concrete WF is related to a large proportion of aggregate (around 5 times) compared with the cement proportion in the concrete production. Recycling aggregates might be a solution to reduce water related aggregates; but the support of this might be a subject of the future research. Based on the results, presented in Fig. 11, the direct water consumption intensity of the selected concrete plant equals 0.22 m³/ton. This is consistent with the findings of Lafarge (2012), Cemex (2015) and Holcim (2015) which give the figures of 0.2–0.25 m³/ton.

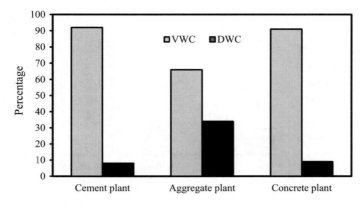

Fig. 10 Proportion of virtual and direct water consumptions for different plants

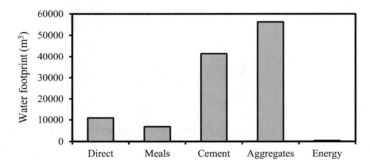

Fig. 11 WF of different parameters in concrete plant

To investigate the role of parameters affecting the WF of cement production Fig. 12 is presented. This figure shows that nature gas and heavy fuel contribute to high proportions (40.18 and 27.12%, respectively) of the total WF of the cement production which are followed by meals, electricity and direct water consumption.

Among three main fossil fuels (crude oil, coal and natural gas), the WF of the natural gas (241 m^3/Tj) is the lowest which is followed by coal (495 m^3/Tj) and crude oil (497 m^3/Tj) (Mekonnen et al. 2015). The WF of refining of natural gas can be shown to be significantly lower than that of the coal and crude oil (Mekonnen et al. 2015). In the case study cement plant, natural gas is used as the main energy source. Thus use of natural gas in cement plants in Iran is in the environmentally friendly manner. The VWC of electricity is close to the DWC in the selected cement plant. By using renewable energy sources for electricity generation the VWC of electricity can be reduced considerably. Mekonnen et al. (2015) demonstrate that the best renewable energy type in terms of WF is wind power with 4.68 L/MW (1.3 m^3/Tj) water consumption intensity compared with 889 L/MW (247 m^3/Tj) of natural gas. However, it might not difficult to make electricity from the wind power

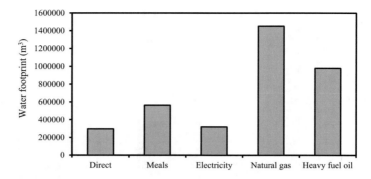

Fig. 12 WF of different parameters in cement plant

Fig. 13 WF of different
parameters in aggregate plant

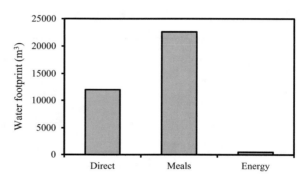

in any part of a nation like Iran. Making electricity from solar is more feasible in Iran due to its weather conditions. Mekonnen et al. (2015) show that the WF of the electricity generation from solar is equal to 504 L/MWh which is lower than that of the nature gas (889 L/MW).

Figure 13 presents the role of different parameters on the WF of aggregate production. This figure demonstrates a high contribution (22,610 m^3) of the personnel' food on the WF of aggregate production which is followed by the direct water and energy consumptions. The food required by personnel, similar to the energy needed by equipment, is a water consumer parameter. Reduction of the personnel number, via for instance using automation tools or upgrading the production technology, can decrease the virtual WF of the personnel significantly. However, this is likely to result in an increase in the WF of energy sources as machines take human jobs.

Based on Fig. 11, the contributions of direct water, energy, personnel food, cement and aggregates on the WF of one cubic meter of concrete production in the concrete plant can be obtained (9.38, 0.36, 5.9, 35.72 and 48.64%, respectively). Also based on Fig. 12 (Fig. 13) the contributions of direct water, energy and personnel food on the WF of one ton of cement (aggregate) production can be calculated which are equal to 8.22 (34.19), 76.15 (1.39) and 15.63 (64.42) percent, respectively. There directly follows that the total contributions of direct water, energy and personnel food (in

	Concrete plant		Cement plant		Aggregates		
Energy	0.36 %	+	0.3572 ×76.15 %	+	0.4864 ×1.39	=	28.24 %
Personnel food	5.90 %	+	0.3572 ×15.63 %	+	0.4864 ×64.42 %	=	42.83 %
Direct water	9.36 %	+	0.3572 ×8.22 %	+	0.4864 ×34.19 %	=	28.93 %

Fig. 14 Contribution of different parameter on WF of concrete production

all plants), on the WF of one cubic meter of concrete production can be determined (28.24, 42.83 and 28.93%, respectively). Figure 14 summarizes the corresponding results. This figure demonstrates that personnel food has the highest impact on the WF of concrete production which is followed by direct water consumption and energy usage. This highlights the role of human management and energy efficiency measures on the WF reduction of the concrete industry.

7 Sustainability Assessment

A sustainability assessment might be performed to evaluate how the water footprints are related to the available water resources in a region. For such assessment water stress can be calculated in a considered region. Water stress is a measurement to evaluate the vulnerability of a nation's water resources and is based on all available water resources and the total amount of water withdrawals. It accounts for the water essential for sustain aquatic ecosystem and for human needs.

In this chapter, a water stress index (WSI) is obtained in the national level of Iran, following a method developed by Smakhtin et al. (2004),

$$WSI = \frac{W}{MAR - EWR} \tag{5}$$

where W is the total fresh water withdrawals, by the humans and the nature; MAR represents the mean annual runoff (total amount of available fresh water) and EWR is the environmental water requirements covering a certain amount of water needed to sustain healthy ecosystems and the benefits they provide.

Water stress is a measurement to evaluate the vulnerability of a nation's water resources and is based on all available water resources and the total amount of water withdrawals. Water stress is similar to water scarcity, which is a measurement of a nations or a basins water supply in relation to the water demand.

Different categories might be adopted to evaluate whether an area or a nation are exposed to intense water stress or not. Smakhtin et al. (2004) provide a category of water stress as demonstrated in Table 5.

Table 5 Water stress categories (Smakhtin et al. 2004)

WSI category	Definition
Less than 0.3	Slightly exploited
Between 0.3 and 0.6	Moderately exploited
Between 0.6 and 1	Heavily exploited
Bigger than 1	Over exploited

FAO (2015) reports MAR and W for Iran 137×10^9 m^3/year and 6.2×10^9 m^3/year, respectively. Following the research of Smakhtin et al. (2004), the EWR needed to maintain healthy aquatic ecosystems differs worldwide from 20 to 50% of average yearly river flow in a basin. In comparison, Herath et al. (2011a) claim an EWR of 30% for all areas of New Zealand while they assigned an EWR of 50% for Netherlands, Germany, Denmark and Sweden. There is no study concerning EWR of Iran. So following Smakhtin et al. (2004), EWR is considered 35% of MAR. There follows by using Eq. (5) the WSI for Iran is equal to 0.71. According to Table 5, the water stress in Iran is higher than 0.6 and therefore Iran is categorized as a heavily exploited region. Iran industrial water withdrawal is 1.1×10^9 m^3/year and its total water withdrawal is 93.3×10^9 m^3/year. Iran's annual concrete production is estimated to be 100 million m^3 (Bod 2014). There follows the WF of concrete production in this country is expected approximately to be 0.23×10^9 m^3 which accounts for 21.1 present of the industrial water withdrawal and 0.25% of the total water withdrawals in Iran. Compared with this result, Gu et al. (2015) demonstrate that the iron production of China stands for around 0.4% of the total water withdrawal of China. It seems that WF of concrete production is high in comparison with other products. As concrete is of particular importance for construction activities, reduction of WF of concrete production should significantly help in maintaining the sustainability of the construction activities globally. Considering the fact that Iran is categorized as a heavily exploited region in the world, such findings show the importance of an effective management strategy in reducing the WF of concrete production which seems vital for water resources and environment. This confirms the necessity of this research to assess the WF of cement, aggregate and concrete plants.

Iran WF research to date have mainly concerned about agricultural industry (Ababaei and Etedali 2014, 2017; Karandish et al. 2015). Fortunately, concerns about industrial WFs and their potential impact on environment have recently taken more attention of researchers (Hosseinian and Nezamoleslami 2018). Using a case study from the concrete industry, as opposed to the agricultural sector, this chapter extends the WF assessment in Iran. The chapter also helps decision makers within the concrete industry in Iran to think about WF of concrete production and its adverse environmental impacts.

8 Sensitivity Analysis

It is acknowledged that it might not be practical to thoroughly assess variability and uncertainties in WF calculations, as the majority of the data lack the information required for such studies (Mack-Vergara and John 2017). Therefore, it is of particular interest to evaluate the sensitivity of the WF amounts among different parameters involved in the WF calculations. This might be obtained via implementing a sensitivity analysis of the WF of the plants by considering the involved parameters (Saltelli et al. 2008). Accordingly, a parametric sensitivity factor approach is utilized (Ditlevsen and Madsen 1996).

The corresponding results are illustrated in Figs. 15, 16 and 17, for the cement, aggregate and concrete plants, respectively.

As anticipated, the relative WF amount, denoted by ΔWCF/WCF, decreases with a decrease of relative such parameters amounts, denoted by ΔP/P. For the cement plant, Fig. 15 demonstrates that a 10% decrease in natural gas, heavy fuel oil, meals, electricity and direct water consummation amounts results in 4.02, 2.71, 1.56, 0.89 and 0.82%, decrease in their relative WF amounts, respectively. This implies that between these five parameters, natural gas is the most sensitive parameter in calculating the WF of the selected cement plant followed by heavy fuel oil, personnel food, electricity and direct water consummation. Similarly, for the aggregate plant, Fig. 16 shows that a 10% decrease in the amounts of personnel foods, direct water consumption, diesel fuel and electricity results in decreasing, ΔWCF/WCF approximately by 6.44, 3.42, 0.09 and 0.05%, respectively. This reveals that between such parameters, personnel food is the most sensitive one in calculating the WF of the

Fig. 15 Sensitivity analysis for cement plant WF parameters

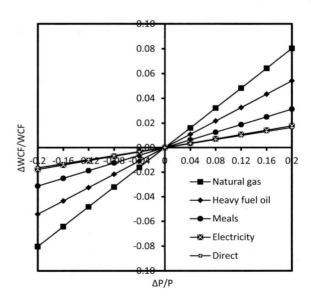

Fig. 16 Sensitivity analysis for aggregate plant WF parameters

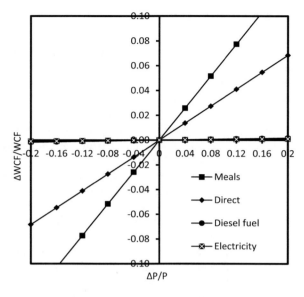

Fig. 17 Sensitivity analysis for concrete plant WF parameters

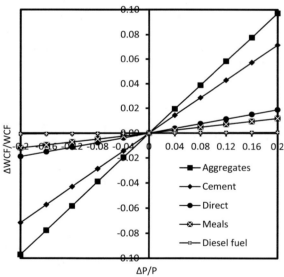

aggregate plant followed by direct water consumption, diesel fuel and electricity. For the concrete plant, Fig. 17 shows that the most sensitive parameter is aggregate which is followed by cement, direct water consumption, personnel food and diesel fuel.

9 Model Validation

Past research mostly focused on the direct water consumption and gave the direct water intensity of concrete plants as 0.2 m^3 per each cubic meter of concrete production (Lafarge 2012; Cemex 2015; Holcim 2015). This chapter demonstrates that by adopting the WF models the WF intensity of the concrete production is equivalent to 2.321 m^3/m^3 which is nearly 11.6 times larger than the direct water intensity of the concrete production. Gu et al. (2015) investigated WF of an iron plant and gave a figure of 0.435 m^3 direct water consumption per each ton of iron. They showed that the WF intensity rises to 5.471 by considering the virtual water which is 12.5 times larger than that of the direct water use. This figure is near to the result of this chapter (11.6 times). In 2014, Netz and Sundin (2015) estimated the WF of 1 ton of concrete 987 L in Sweden. Considering the density of 2.40 ton/m^3 for concrete the WF intensity of 1 m^3 of concrete in Sweden is 2.35 m^3/m^3 which is close to this research results.

10 Conclusion

This chapter made a genuine contribution by empirically exploring WF of concrete industry based on a life cycle assessment approach. A comprehensive WF model of concrete production was proposed by considering the type of energy consumption, transportation and human effects. To provide a metric for WF of concrete production three plants, cement, aggregate and concrete, were analyzed. The chapter shows that the WF of the case study cement, aggregate and concrete plants with the current production rate (1.7 million tonnes cement, 60,000 tonnes aggregates and 50,000 m^3 concrete per year) contributes to 3.614 million m^3, 35,000 m^3 and 116,064 m^3, with the water consumption intensity of 2.126, 0.583 and 0.967 m^3/ton (2.321 m^3/m^3), respectively, in 2017. The chapter demonstrates a large contribution of concrete production to water footprint and risk of such production in dry countries. The chapter also demonstrates that in the selected cement, aggregate and concrete plants virtual WFs are 11, 1.9, 11.6 times larger than those of direct water consumptions. Additionally, the chapter shows that most of the water use of the cement plant is associated with the virtual water of the fossil energy consumption (2752,000 m^3). The role of energy efficiency actions and utilizing renewable energy sources in reducing WF is central. Furthermore, the chapter illustrates that the personnel's food contributes to 6850, 565,000, and 22,610 m^3, in the concrete, cement and aggregate plants, respectively; illustrating the role of human management on the WF reduction.

The chapter findings allow the concrete industry to take measures for reducing the related environmental influences of concrete life cycle. By providing benchmarks for improvement, the chapter findings help in underpinning strategies for reducing pressure on water resources. This chapter assists in understanding the WF methodology which results in more detailed and clarified impact assessment of concrete life cycle on environment management.

The WF assessment was conducted on three sample plants. Further empirical investigation needs to be carried out on different concrete, aggregate and cement plants to give further support for the chapter's findings.

References

Ababaei, B., & Etedali, H. R. (2014). Estimation of water footprint components of Iran's wheat production: Comparison of global and national scale estimates. *Environmental processes, 1*(3), 193–205.

Ababaei, B., & Etedali, H. R. (2017). Water footprint assessment of main cereals in Iran. *Agricultural Water Management, 179,* 401–411.

Acquaye, A., Feng, K., Oppon, E., Salhi, S., Ibn-Mohammed, T., Genovese, A., et al. (2017). Measuring the environmental sustainability performance of global supply chains: A multi-regional input-output analysis for carbon, sulphur oxide and water footprints. *Journal of Environmental Management, 187,* 571–585.

Allan, J. (1998). Virtual water: A strategic resource global solutions to regional deficits. *Journal of groundwater, 36,* 545–546.

Bod, A. M. (2014). An analysis of cement industry. *Cement Technology., 73,* 120. (in Farsi).

Bruinsma, J. (2009). *The resource outlook to 2050: By how much do land, water, and crop yields need to increase by 2050?* FAO Expert Meeting on 'How to feed the world in 2050'. Retrieved from Rome.

Carmichael, D.G., & Balatbat, M.C. (2009). Sustainability on construction projects as a business opportunity. In *SSEE 2009: Solutions for a sustainable planet* (pp. 465–474). Barton, A.C.T.: Engineers Australia.

Cemex. (2015). *Sustainable Development Report*. Building Resilient and Sustainable Urban Communities.

Chapagain, A. K., & Orr, S. (2009). An improved water footprint methodology linking global consumption to local water resources: A case of Spanish tomatoes. *Journal of Environmental Management, 90,* 1219–1228.

Chehreghni, H. (2004). *Environment in cement industry*. Hazegh publications (in Farsi).

Chen, C., Habert, G., Bouzidi, Y., & Jullien, A. (2010). Environmental impact of cement production: Detail of the different processes and cement plant variability evaluation. *Journal of Cleaner Production, 18,* 478–485.

Chico, D., Aldaya, M. M., & Garrido, A. (2013). A water footprint assessment of a pair of jeans: The influence of agricultural policies on the sustainability of consumer products. *Journal of Cleaner Production, 57,* 238–248.

Ditlevsen, O., & Madsen, H. O. (1996). *Structural reliability methods*. Hoboken, NJ: Wiley.

Doe, U. (2006). *Energy demands on water resources. Report to congress on the interdependency of energy and water. Washington DC: US Department of Energy, 1.*

Domenech Quesada, J. L. (2007). *Huella Ecológica y Desarrollo Sostenible (Ecological Footprint and Sustainable Development).* Madrid, Spain: AENOR.

Enerdata. Global energy and CO_2 data. www.enerdata.net. Accessed June 26, 2015.

Ercin, A. E., Aldaya, M. M., & Hoekstra, A. Y. (2011). Corporate water footprint accounting and impact assessment: The case of the water footprint of a sugar-containing carbonated beverage. *Journal of water resources management, 25*(2), 721–741.

Ercin, A. E., Aldaya, M. M., & Hoekstra, A. Y. (2012). The water footprint of soy milk and soy burger and equivalent animal products. *Journal of ecological indicators, 18,* 392–402.

FAO. (2015). *AQUASTAT Database*. Available from: http://www.fao.org/nr/water/aquastat/main/index.stm, visited March, 2018.

Finnveden, G., Hauschild, M. Z., Ekvall, T., Guinee, J., Heijungs, R., Hellweg, S., et al. (2009). Recent developments in life cycle assessment. *Journal of Environmental Management, 91*(1), 1–21.

Finnveden, G., & Moberg, A. (2005). Environmental systems analysis tools—an overview. *Journal of Cleaner Production, 13,* 1165–1173.

Gao, C., Wang, D., Dong, H., Cai, J., Zhu, W., & Du, T. (2011). Optimization and evaluation of steel industry's water-use system. *Journal of Cleaner Production, 19*(1), 64–69.

Gartner, E. (2004). Industrially interesting approaches to "low-CO_2" cements. *Cement and Concrete Research, 34,* 1489–1498.

Georgiopoulou, M., & Lyberatos, G. (2017). Life cycle assessment of the use of alternative fuels in cement kilns: A case study. *Journal of Environmental Management*.

Gu, Y., Xu, J., Keller, A. A., Yuan, D., Li, Y., Zhang, B., et al. (2015). Calculation of water footprint of the iron and steel industry: a case study in Eastern China. *Journal of Cleaner Production, 92,* 274–281.

Hasanbeigi, A., Price, L., & Lin, E. (2012). Emerging energy-efficiency and CO_2 emission-reduction technologies for cement and concrete production: a technical review. *Renewable and Sustainable Energy Reviews, 16,* 6220–6238.

Herath, I., Clothier, B., Horne, D., & Singh, R. (2011a). *Assessing freshwater scarcity in New Zealand*. New Zealand Life Cycle Management Centre Working paper Series 02/11, pp. 21–26.

Herath, I., Deurer, M., Horne, D., Singh, R., & Clothier, B. (2011b). The water footprint of hydro-electricity: A methodological comparison from a case study in New Zealand. *Journal of Cleaner Production, 19*(14), 1582–1589.

Hoekstra, A. Y. (2002). Virtual water trade: Proceedings of the international expert meeting on virtual water trade. In A. Y. Hoekstra (Ed.), *Value of Water Research Report*. The Netherlands: IHE Delft.

Hoekstra, A.Y. (2008). The water footprint of food. In Forare, J. (ed.) *Water for food* (vol. 28, pp. 49–60). The Swedish Research Council for Environment, Agricultural Sciences and Spatial Planning (Formas), Stockholm, Sweden.

Hoekstra, A. Y., Chapagain, A. K., Aldaya, M. M., & Mekonnen, M. M. (2011). *The water footprint assessment manual. Setting the global standard*. London: Earthscan.

Hoekstra, A. Y., & Mekonnen, M. M. (2012). The water footprint of humanity. *Proceedings of the National Academy of Sciences, 109,* 3232–3237.

Holcim. (2015). *Corporate Sustainable Development Report*. Building on Ambition.

Horvath, A., & Matthews, H. S. (2004). Advancing sustainable development of infrastructure systems. *Journal of Infrastructure Systems, 10*(3), 77–78.

Hosseinian, S. M., & Nezamoleslami, R. (2018). Water footprint and virtual water assessment in cement industry: A case study in Iran. *Journal of Cleaner Production, 172,* 2454–2463.

Hunt, R. G., Sellers, J. D., & Franklin, W. E. (1992). Resource and environmental profile analysis: A life cycle environmental assessment for products and procedures. *Environmental Impact Assessment Reviews, 12,* 245–269.

Huntzinger, D. N., & Eatmon, T. D. (2009). A life-cycle assessment of Portland cement manufacturing: comparing the traditional process with alternative technologies. *Journal of Cleaner Production, 17,* 668–675.

International Organization for Standardization (ISO). (2006a). *ISO 14040:2006, Environmental management—Life cycle assessment—Principles and framework*.

International Organization for Standardization (ISO). (2006b). *ISO 14044:2006, Environmental management—Life cycle assessment—Requirements and guidelines*.

International Organization for Standardization (ISO). (2014). *ISO 14046:2014 Environmental Management—Water footprint—Principles, requirements and guidelines*.

Jeswani, H. K., & Azapagic, A. (2011). Water footprint: methodologies and a case study for assessing the impacts of water use. *Journal of Cleaner Production, 19,* 1288–1299.

Josa, A., Aguado, A., Cardim, A., & Byars, E. (2007). Comparative analysis of the life cycle impact assessment of available cement inventories in the EU. *Cement and Concrete Research, 37*(5), 781–788.

Josa, A., Aguado, A., Heino, A., Byars, E., & Cardim, A. (2004). Comparative analysis of available life cycle inventories of cement in the EU. *Cement and Concrete Research, 34*(8), 1313–1320.

Karandish, F., Salari, S., & Darzi-Naftchali, A. (2015). Application of virtual water trade to evaluate cropping pattern in arid regions. *Water Resources Management, 29*(11), 4061–4074.

Kibert, C. J., & Fard, M. M. (2012). Differentiating among low-energy, low-carbon and net-zero-energy building strategies for policy formulation. *Journal of Building Research & Information, 40*(5), 625–637.

King, C. W., & Webber, M. E. (2008). Water intensity of transportation. *Environmental Science & Technology, 42*(21), 7866–7872. ACS Publications.

Knoema. (2005). *National water footprint statistics.* https://knoema.com/WFPNWFPS2015/nation al-water-footprint-statistics-1996-2005?location=1000910-iran. Accessed September 10, 2017.

Korre, A., & Durucan, S. (2009). *Life cycle assessment of aggregates.* EVA025—Final Report: Aggregates industry life cycle assessment model: modelling tools and case studies published by WRAP.

Lafarge. (2012). Sustainability 11th Report 2011.

Mack-Vergara, Y. L., & John, V. M. (2017). Life cycle water inventory in concrete production—A review. *Journal of resources, conservation and recycling, 122,* 227–250.

Mekonnen, M.M., Gerbens-Leenes, P., & Hoekstra, A.Y. (2015). The consumptive water footprint of electricity and heat: A global assessment. Environmental science. *Journal of Water Research & Technology, 1*(3), 285–297.

Mekonnen, M. M., & Hoekstra, A. Y. (2014). water footprint benchmarks for crop production: A first global assessment. *Journal of ecological indicators, 46,* 214–223.

Mellor, A.E. (2017). *Assessing water footprint and associated water scarcity indicators at different spatial scales: a case study of concrete manufacture in New Zealand: a thesis presented in partial fulfilment of the requirements for the degree of Master in Environmental Management*, Massey University, Manawatu Campus, New Zealand (Doctoral dissertation, Massey University).

Mielke, E., Anadon, L.D., & Narayanamurti, V. (2010). *Water consumption of energy resource extraction, processing, and conversion.* Belfer Center for Science and International Affairs.

Ness, B., Urbel-Piirsalu, E., Anderberg, S., & Olsson, L. (2007). Categorising tools for sustainability assessment. *Journal of ecological economics, 60*(3), 498–508.

Netz, J., & Sundin, J. (2015). *Water footprint of concrete.* (Environmental Strategies, Second Cycle), Royal Institute of Technology.

Neville, Adam M. (1995). *Properties of concrete* (Vol. 4). London: Longman.

Owens, J. W. (1997). Life cycle assessment: constraints on moving from inventory to impact assessment. *Journal of Industrial Ecology, 1*(1), 37–49.

Pfister, S., Saner, D., & Koehler, A. (2011). The environmental relevance of freshwater consumption in global power production. *The International Journal of Life Cycle Assessment, 16*(6), 580–591.

Ridoutt, B. G., & Pfister, S. (2010). Reducing humanity's water footprint. *Journal of Environmental Science Technology, 44*(16), 6019–6021.

Sakai, K., & Noguchi, T. (2013). *The sustainable use of concrete.* Boca Raton: CRC Press.

Saltelli, A., Ratto, M., Andres, T., Campolongo, F., Cariboni, J., Gatelli, D., et al. (2008). *Global sensitivity analysis: The primer.* Hoboken, NJ: Wiley.

Schneider, M., Romer, M., Tschudin, M., & Bolio, H. (2011). Sustainable cement production present and future. *Cement and Concrete Research, 41*(7), 642–650.

Schulte, P. (2014). *Defining water scarcity, water stress, and water risk: It is not just semantics.* Retrieved from http://pacinst.org/water-definitions/.

Scown, C. D., Horvath, A., & McKone, T. E. (2011). Water footprint of U.S. transportation fuels. *Environmental Science and Technology, 45,* 2541–2553.

Shiklomanov, I. A., & Rodda, J. C. (2003). *World water resources at the beginning of the 21st century.* Cambridge, UK: UNESCO and Cambridge University Press.

Siew, R. A. (2015). Review of corporate sustainability reporting tools (SRTs). *Journal of Environmental Management, 164,* 180–195.

Sjunnesson, J. (2005). *Life cycle assessment of concrete.* Master thesis, Lund University, Sweden.

Smakhtin, V., Revenga, C., & Döll, P. (2004b). *Taking into account environmental water requirements in the global-scale water resource assessment.* Retrieved from Colombo, Sri Lanka.

Spiess, W. (2014). Virtual water and water footprint of food production and processing. In *Encyclopedia of agriculture and food systems* (pp. 333–355).

United Nations. (2013). *World populations' prospects: The 2012 revision, Volume 1: Comprehensive tables.* Retrieved from New York, USA.

Valderrama, C., Granados, R., Cortina, J. L., Gasol, C. M., Guillem, M., & Josa, A. (2012). Implementation of best available techniques in cement manufacturing: A life-cycle assessment study. *Journal of Cleaner Production, 25,* 60–67.

Van Oel, P., & Hoekstra, A. (2012). Towards quantification of the water footprint of paper: A first estimate of its consumptive component. *Journal of water resources management, 26,* 733–749.

Verma, S., Kampman, D. A., Van Der Zaag, P., & Hoekstra, A. Y. (2009). Going against the flow: A critical analysis of inter-state virtual water trade in the context of India's National River Linking Program. *Journal of physics and chemistry of the earth, Parts A/B/C, 34*(4), 261–269.

Williams, E.D., & Simmons, J.E. (2013). *Water in the energy industry: An introduction.* www.bp.com/energysustainabilitychallenge, visited on June 2017.

World Business Council for Sustainable Development. (2014). *Protocol for water reporting.*

Zhang, G., Sandanayake, M., Setunge, S., Li, C., & Fang, J. (2017). Selection of emission factor standards for estimating emissions from diesel construction equipment in building construction in the Australian context. *Journal of Environmental Management, 187,* 527–536.

Zhuo, L., Mekonnen, M. M., & Hoekstra, A. Y. (2016). The effect of inter-annual variability of consumption, production, trade and climate on crop-related green and blue water footprints and inter-regional virtual water trade: A study for China (1978–2008). *Journal of water research, 94,* 73–85.

Printed in the United States
By Bookmasters